T0073039

Environmental Footprints and Eco-design of Products and Processes

Series editor

Subramanian Senthilkannan Muthu, SgT Group and API,
Hong Kong, Hong Kong

This series aims to broadly cover all the aspects related to environmental assessment of products, development of environmental and ecological indicators and eco-design of various products and processes. Below are the areas fall under the aims and scope of this series, but not limited to: Environmental Life Cycle Assessment; Social Life Cycle Assessment; Organizational and Product Carbon Footprints; Ecological, Energy and Water Footprints; Life cycle costing; Environmental and sustainable indicators; Environmental impact assessment methods and tools; Eco-design (sustainable design) aspects and tools; Biodegradation studies; Recycling; Solid waste management; Environmental and social audits; Green Purchasing and tools; Product environmental footprints; Environmental management standards and regulations; Eco-labels; Green Claims and green washing; Assessment of sustainability aspects.

More information about this series at http://www.springer.com/series/13340

Subramanian Senthilkannan Muthu
Editor

Environmental Water Footprints

Agricultural and Consumer Products

Editor
Subramanian Senthilkannan Muthu
SgT Group and API
Hong Kong, Kowloon
Hong Kong

ISSN 2345-7651 ISSN 2345-766X (electronic)
Environmental Footprints and Eco-design of Products and Processes
ISBN 978-981-13-2507-6 ISBN 978-981-13-2508-3 (eBook)
https://doi.org/10.1007/978-981-13-2508-3

Library of Congress Control Number: 2018953647

This Springer imprint is published by the registered company Springer Nature Singapore Pte Ltd.
The registered company address is: 152 Beach Road, #21-01/04 Gateway East, Singapore 189721, Singapore

This book is dedicated to:
The lotus feet of my beloved
Lord Pazhaniandavar
My beloved late Father
My beloved Mother
My beloved Wife Karpagam and
Daughters—Anu and Karthika
My beloved Brother
Everyone working in the Agricultural
and Consumer Products sectors to make it
ENVIRONMENTALLY SUSTAINABLE

Contents

Water Footprint of Agricultural Products

P. Senthil Kumar and G. Janet Joshiba

Abstract Water is a greatest gift of nature and it is an essential necessity of all living organism to survive in earth. Due to the over exposure of industrialization and urbanization the quantity and the quality of the water sources are getting depleted, thus it is necessary to measure the volume of the water consumed to create each of the merchandise and enterprises we utilize. The water foot printing helps people, organizations and nations by disclosing the amount of water utilized by all sectors from an individual level to national level. They also help in highlighting the volume of water utilized as a part of the considerable number of procedures engaged with assembling and delivering our products. They likewise represent the measure of water defiled amid assembling and generation. Agriculture is one of the major consumers of water and it fully depends on water for growth and production of agricultural products. It is directly and indirectly linked with the water scarcity, Furthermore it is also one of the major sources affecting the quality of water by overloading the fresh water with excess amount of nutrients. Food production is directly linked with the water scarcity and so to increase the food production it is essential to measure the water utility in agricultural process. Water footprint of agricultural products aims in developing new strategies to overcome the water scarcity in agricultural sector and it helps in administering many effective rules for empowering a more proficient governance of water sources under climatic changes, industrial pollution and various other factors affecting the quality of water. This review narrates the various features of the water footprint in the agricultural products.

Keywords Water footprint · Water scarcity · Food production · Agriculture

P. Senthil Kumar (✉) · G. Janet Joshiba
Department of Chemical Engineering, SSN College of Engineering,
Chennai 603110, India
e-mail: senthilchem8582@gmail.com; senthilkumarp@ssn.edu.in

S. S. Muthu (ed.), *Environmental Water Footprints*, Environmental Footprints and Eco-design of Products and Processes,
https://doi.org/10.1007/978-981-13-2508-3_1

1 Introduction

Water is one of the greatest gifts of Mother Nature. Life without water is unimaginable in earth. Water is an essential source for energy and every living being in earth depends on water for energy. Nowadays, due to high population growth, advancement in economy, industrial growth, changing climatic conditions, etc., causes depletion of fresh water sources (Hogeboom et al. 2018). With the explosion of population outgrowth combined with changing eating regimen inclinations, water withdrawals are relied upon to keep on increasing in the coming decades universally, the water scarcity has expanded about seven fold in the previous century. Even after the implementation of wastewater treatment in the industries, the used water cannot be regained for other purpose because of the unpredictable global warming, ozone depletion and climatic changes which are causing evaporation of water sources. Furthermore, around 85% of global water is utilized for farming purposes (Mekonnen and Hoekstra 2011). The developing assemblage of research on water utilize, shortage and contamination in connection to utilization, generation and exchange has prompted the development of the field of Water Footprint (Hoekstra 2017). The water footprinting of items is generally used by the private organisations to perform chance evaluation and it is used as an instrument to distinguish hotspots in their supply affixes or to couple it with instruments like LCA techniques keeping in mind the end goal to perform benchmarking of items. In the second case, it is used to provide the fundamental data of the occupants of a particular nation who are consuming the water sources and it is used by the scholarly community, Non-Governmental Organisations and private sectors (UNEP 2011). Agriculture is the largest global consumer of fresh water sources. The risk of depletion of global fresh water sources increases due to the outgrowth of inevitable environmental issues causing depletion of water sources globally leading to deficiency of fresh water sources for agricultural consumption. The water utilization for crop production is getting elevated every year by 0.7% and the requirement of water for crop production will increase from a level of 6400 Gm^3/yr to 9060 Gm^3/yr by the year 2050 to nourish around 9.2 billion of the global population (Mekonnen and Hoekstra 2013). Ground water is a best source of fresh water which meets the requirements of billions of people and it plays an extraordinary role in the agriculture which remains as the basic food production source of human beings. The ground water level is evaluated by many governmental and non-governmental organizations to know about the rate of depletion of ground water due to some environmental issues (Gleeson et al. 2012). Water acts as an outstanding resource of nature and it place a phenomenal role in the development of merchandise and enterprises, furthermore, the water utilized as a part of the manufacturing sector is known as the virtual water. For manufacturing 1 kg of grain we need approximately 1000–2000 kg of water. For delivering 1 kg of meat we require in normal 16000 kg of water, whereas for producing 1 kg of cheddar we require about 5000–5500 kg of water. Every item manufactured in an industrial sector compulsorily needs water as a main raw material (Hoekstra 2003). Various water consuming activities followed globally are depicted in the (Fig. 1). The developing freshwater shortage is as of now

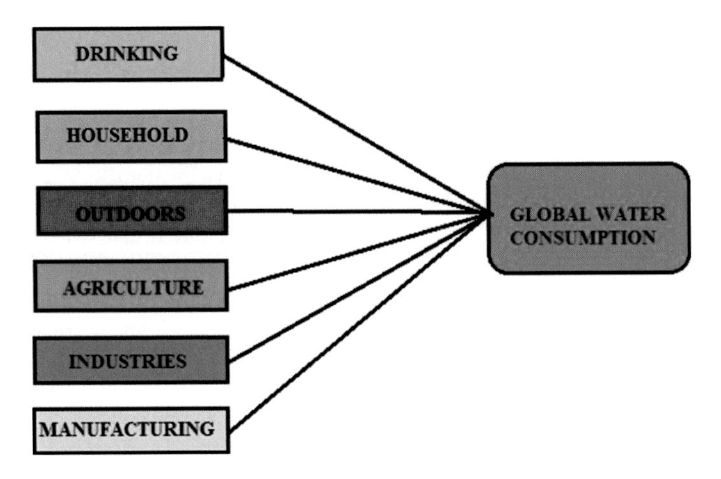

Fig. 1 Various water consuming activities

apparent in numerous parts of the world. Raising water scarcity increases the risk for deficiency of fresh water sources for agricultural utilization which leads to depletion in food production. The water footprint offers a quantifiable marker to quantify the volume of water utilization per unit of harvest, and in addition the volume of water contamination (Mekonnen and Hoekstra 2013). This article elucidate the impact of agriculture on the global water sources and it gives an clear description of the water footprinting concepts, its methodology, data collection method and it also explains the importance of water footprinting on the agricultural products and its impact on developing the global fresh water sources.

2 Global Demand for Water

2.1 Water Footprint Concept

The water footprint concept was enunciated in an international expert meeting on virtual water trade in the year 2002, December. Hoekstra and Hung have enunciated the idea of the water footprinting to know about this overall amount of virtual water substance of all merchandise and enterprises devoured by one individual or by the people of one nation. When compared with the other resources such as land, air and energy, only little research has been done in the water and its impact on its utilization by industries for manufacturing goods and services (Hoekstra 2003). The water footprinting measures the quantity of water used in manufacturing a product into every one of the merchandise and enterprises globally and it also likewise reveals about the amount of water being devoured by every nation globally (Water footprinting network). They presented the water footprint as a marker for utilization of water that behind every products and ventures devoured by one individual or the people of a nation a they also assured it as a good tracking system of a nation to know about its water resources. Water footprints in simple terms can be used to track the individual impression as a component of nourishment eating routine and utilization pattern (Hoekstra 2017). The water footprint inspects both the direct and indirect use of water for a manufacturing process. It is likewise conceivable to utilize the water footprinting to gauge the measure of water required to create every one of the merchandise and ventures devoured by the individual or group, a country or all of humankind. This likewise incorporates the immediate water impression, which is the water utilized specifically by the individual and the aberrant water impression which is the summation of the water impressions of the considerable number of items devoured. It also peeks into the amount of water utilized for the manufacturing a product right from the beginning to the end of the manufacturing process (Water footprinting network). "Virtual water content" is the another terminology given to the water footprint of a product. On account of agricultural products, the water footprinting is quantified in terms of m^3/ton or liters/kg. The water used for agriculture can also be calculated in terms of the volume of water utilized to produce per product (Hoekstra et al. 2011). The water footprint can be viewed as an extensive marker of new water assets allocation, alongside the conventional and limited measures of water withdrawal. It gives information about the following data:

- Type of water utilized: Blue (identified with new surface or ground water), Green (related precipitation put away in the dirt as soil dampness) and Gray Water (identified with water contamination).
- Virtual Water Flows: The virtual water stream between two topographically portrayed zones is the volume of virtual water that is being exchanged from one to alternate because of item exchange.
- Spatial and Temporal localisation of the Water Footprints: all parts of an aggregate Water Footprint are determined topographically and transiently (UNEP 2011).

3 Three Water Footprints

3.1 Green Water Footprint

Green water footprint is measure of water from precipitation that is reserved in the root zone of the soil and dissipated, absorbed and consolidated by the plants. It is applicable to farming, forestry and horticulture.

3.2 Blue Water Footprint

Blue water footprint is the measure of water that has been locked up in the lower surface of the earth. It basically takes a measure of ground water taken from one waterway and come back to another, or returned at an alternate time. Inundated agribusiness, industry and household water utilize can each have a blue water footprinting.

3.3 Grey Water Footprint

Grey water footprint is the measure of the water required to absorb contaminants to meet particular water quality norms. The grey water considers point-source contamination released to a freshwater asset specifically through a pipe or in a roundabout way through overflow or filtering from the dirt, impenetrable surfaces, or other diffuse sources (Mekonnan and Hoekstra 2012). The various water fooprints and their specific function are depicted in Fig. 2.

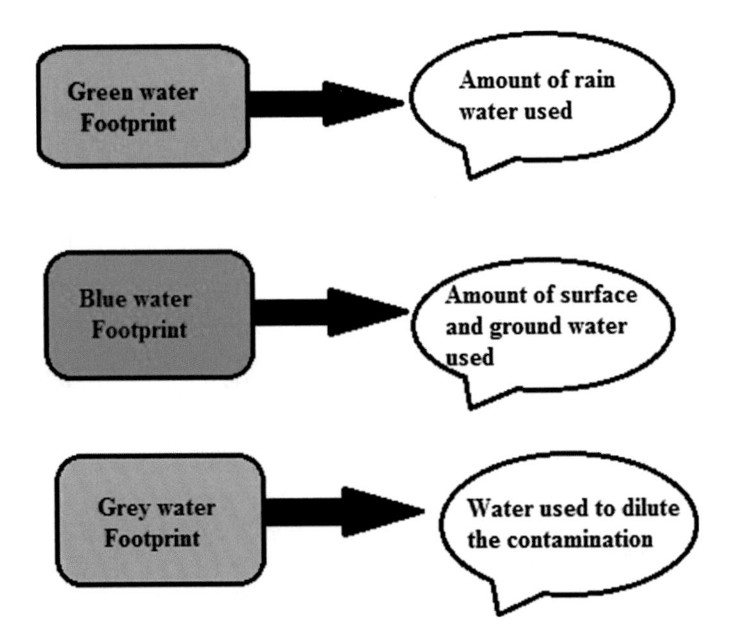

Fig. 2 Different water footprints and their function

4 Water Footprint Assessment

Water footprint assessment quantifies the quantity of available fresh water and it segregates the water into green, blue and grey water footprints. It helps in surveying the manageability, effectiveness and even-handedness of utilization of water for various industrial and household activities. Water Footprint Assessment is flexible and can illuminate a wide scope of key activities and approaches from natural, social and financial points of view. The water footprinting offers an extensive relationship between the consumer and the volume of water utilized to manufacture a product. It is a volumetric measure of water utilization and contamination. Water footprint assessment briefly explains the relationship of water with the human purposes. They can bolster the talk about supportable and fair water utilize and allotment and furthermore frame a decent reason for a nearby appraisal of natural, social and financial effects (Water footprint network). The procedure isn't intended to be direct or a strict order, yet it gives a sorted out system. Before the studying the various parameters involved in the utilization of water, it is important to understand the purpose of the examination. This assessment is used by the governmental and non-governmental organizations to understand the water availability and the utilization of water in producing various items (UNEP 2011). Water footprinting of agricultural products aids in determining the consumption of water sources in various regions and comparing the water utilization rate of different domains. It helps in finding out the factors responsible for reducing the consumption of fresh water sources and helps in building new strategies and ideas for increasing the fresh water sources. The water footprinting assessment also extends its view towards the contamination caused due to the crop production and its harmful impacts on water sources. The water footprinting of agricultural products differs based on the different geographical locations and environmental conditions (Mekonnen and Hoekstra 2013).

5 Four Sectors of Water Footprinting Assessment

A Water Footprint Assessment starts with defining the objectives and scope of the study and it can be used for various purposes. It is used in enhancing the management of utilization of water in various organizations, industrial sectors, Private sectors, etc. It is used in achieving the management of water sources utilized in direct and indirect way. The water footprint assessment helps in developing the benchmarks for utilization of water sources and contamination of water sources caused by the industries during the manufacture of a product. This creates awareness about the management of water sources in various domains.

5.1 Goals and Scope

A Water Footprint Assessment can be custom fitted to meet the objectives and extent of the investigation. The goal and scope of the water footprint plans and explains the steps to be undertaken in the following phases such as accountability, Sustainability assessment and resource formulation. The extent of the scope characterizes the size and the type of investigation to be undertaken in the water footprinting of the organization. The goal and scope of the water footprinting assessment is used to demonstrate the type of information utilized for the study and to derive the various strategies for lowering the water footprint.

5.2 Accountability

In the accountability phase of water footprinting assessment, information is gathered and accounts are created relying upon the purpose of the utilization. Once the objective and extent of the Water Footprint Assessment have been characterized, the information is gathered to ascertain the impression of the applicable procedures for the investigation. These may originate from worldwide databases, for example, WaterStat, or gathered locally. The figurings for the green, blue and dark water impression take after the procedure portrayed in the Water Footprint Assessment Manual.

5.3 Sustainability Assessment

In the sustainability assessment phase of the water footprinting assessment, the utilization of water sources is evaluated from social, environmental and economic viewpoint. Water Footprint Assessment is utilized to evaluate whether water utilize is naturally maintainable, asset productive and fairly allotted. In this phase, the place is surveyed for proper utilization of water assets in meeting the requirements of the industries and other sectors.

5.4 Response Formulation

Response formulation is the final phase of the water footprinting assessment and in this phase according to the results of the survey various policies, standards and strategies are enunciated. Response formulation helps in gathering the information and giving out correct policies and strategies to empower enhanced water administration, Futhermore it implements some changes in practices and it increases innovation that will diminish the water footprint at any progression along the esteem chain. It might likewise be essential to make a move on the whole with others to enhance the long haul manageability of water use at the catchment or stream bowl level (Hoekstra 2017).

6 Calculation of Water Footprinting

There are at present three methodologies being connected for the figuring of a water impression. They are:

- Volumetric approaches
- Stress weighted approaches
- Life cycle assessment approaches

6.1 Volumetric Approaches

The volumetric approach of the water footprinting fully depends on an evaluation of the volume of water related with a specific generation movement.

6.2 Stress Weighted Approaches

The stress weighted approach depends on an appraisal of the measure of freshwater expended in a generation movement joined with an appraisal of the ramifications of that utilization as far as water pressure.

6.3 Life Cycle Assessment Approaches

Life cycle assessment approach of the water footprinting depends on evaluations of water utilization utilizing a stock examination like that of the volumetric approach yet in addition counting a component of effect evaluation (RPA 2011).

7 Water Consumption on Agricultural Sector

Agriculture is said to be the backbone of our country and it plays an outstanding role in the global food production. The evolution in industrialization, urbanization and other environmental changes will have a negative impact on the agricultural water. It represents around 70% of water utilized as a part of the world today and furthermore adds to contamination from abundance supplements, pesticides and different poisons. Due to advancement of science and technology the level of water contaminants are also elevated and it directly leads to the high expenses on water treatment. The global warming and greenhouse gas emission causes alteration in the regular planning of precipitation and it leads to the melting of polar ice caps, furthermore it also causes seriousness of surges and dry seasons (OECD 2016). Agriculture water is utilized to develop and produce fresh food sources which are the basic need of every human being. As indicated by the United States Geological Survey (USGS), around 65% of the world's fresh water is used for agricultural purpose; furthermore, around 330 million section of land in United States is used to create a plenitude of sustenance and different items. Agricultural water are derived from different sources such as rivers, lakes, streams, ditches, canals, ponds, reservoirs, wells and rainwater, etc. Water form these sources are used for cultivation purpose (CDC 2016). The amplification of the agricultural zone leads to outflows of ozone harming substances which interrupts in the living of the biological systems (Chukkala et al. 2018). In the total water footprinting of agriculture about 29% of the water is contributed to the farming division and about 33% of the total global water footprint is involved with farming of cattle and sheep's. The blue and grey water footprints of the global water footprinting reveals that the water used for animal production is more than the water used for agricultural crop production (Mekonnen and Hoekstra 2012). The reasonable administration of water in agriculture is affected due to the urbanization and other technological advancement. Governments need to enhance the financial proficiency and ecological adequacies of arrangements that try to enhance water asset utilize effectiveness and decrease water contamination from farming frameworks (OECD 2016). The monetary estimation of water utilized for agricultural purposes is calculated by duplicating the inundated zone overhauled by the reservoir with the normal financial estimation of rural land in the nation where the repository is found. The normal monetary estimation of agric-social land is evaluated per nation by separating the estimation of farming generation of all yields in the nation by the manufacturing zone of all products in the nation. In the year 2013, the water footprinting for agricultural products produced per crop per nation were derived from the FAOSTAT (Hogeboom et al. 2018).

8 Need for Water Footprinting in Agriculture

The water footprinting is a developing strategy to study about the water consumption and distribution. It is a fundamental tool to study about the amount of water utilized to manufacture a product. The water footprinting is mainly required to evaluate the indirect use of water, due to its accessibility worldwide. The unsustainable water utilization is progressively determined by the worldwide economy which needs impetuses for manageable water utilize. As the freshwater reestablishment rates are restricted, so we should contemplate the advancement of Utilization, generation and exchange designs in connection to these constraints. In more extensive sense, while breaking down the natural manageability of economies, it is important to ponder the water footprinting of human utilization in connection to planetary limits. The understanding of the regular utilization and consumption level of water source should be made mandatory to improve the supply chains and life cycles. The water footprint is a perfect tool for knowing the fresh water utilization and shortage; furthermore, it is also used to know the extent of water contamination (Hoekstra 2017).

9 Water Footprinting of Agricultural Products

Water is a basic factor in agribusiness and assumes a conclusive part in development and advancement of plants and trees (Juan et al. 1999). Water footprinting of the agricultural activates is an essential framework required by the people to maintain sustainable management of fresh water sources. It helps in demonstrating the manageability of the grain development and it point outs the rivalry for water amongst nourishment and vitality crops. Water Footprinting gives area and time particular data on the difficulties such as blue water utilization and grey water contamination. Water footprinting gives information about the green water utilization openings of the distinctive harvest creation. WF of the diverse frameworks and strategies can be contrasted with illuminate and enhance basic leadership. Water Footprinting can be utilized to make mindfulness on water utilization identified with conduct (UNEP 2011). The development of farming in a worldwide market has demonstrated that water system is a fundamental system for the development of vital condition of agro-nourishment sectors. In this regard, moves to be made must reinforce water system cultivating, ensure the country condition and bolster water asset maintainability with great water administration (Juan et al. 1999). The water footprinting has been proposed as a metric that demonstrates the water utilize and effects of the generation framework on water assets. The water footprinting studies are usually performed using several approaches such as hydrological outflow, hydrological inflow and storage modifications. These methods reveal the influence of water consumption on agricultural crop production (Herath et al. 2013).

10 Methodology and Data

The water footprinting assessment categorizes the water sources based on green, blue and grey footprints. In calculating the water footprint of agricultural products the evaporation and transpiration rates of the crop are required and furthermore, the yield of the crop in that particular region is an important parameter to be estimated in the water footprint of agricultural water framework. The unique grid based models are used to estimate the balance between the soil and water. The water required for the growth of the crops and the amount of the water consumed by the crops to produce that yield is used as the key parameter in calculating the water footprints of the agricultural products. The CROPWAT model is the often used model for assessment of the water footprints of the agricultural products. The effective water consumption of agricultural products mainly depends on the type of the crop cultivated, soil requirement, water consumption and climatic conditions (Mekonnen and Hoekstra 2011). The grey water footprint is determined by calculating the amount of nitrogen used, amount of nitrate mixed in the agricultural runoff and the amount of nitrate leached into the ground causing ground water contamination. It is the measure of the amount of pesticides, manure and other supplements altering the quality of the fresh water (Mekonnen and Hoekstra 2014).

11 Calculation of Water Footprinting of Agriculture

In the beginning the water footprint is evaluated for the crop production as it involves a very large amount of water for farming purposes. Initially, the water footprinting was analysed by Crop Wat model of FAO using the nationwide production statistics and international trade details. This Crop Wat model is used to determine the water footprinting of the crop production. The main worldwide lattice based evaluation, at 5×5 curve minute determination, was distributed in 2011 using CROPWAT model for evaluating Water Footprinting in crop production. Nowadays, Aqua Crop soil-water-balance and crop growth model are used to study the water utilization for crop production in agriculture (Hoekstra 2017). The water footprinting of the agricultural products are mainly divided into two parameters such as water footprinting of production sector (WF_{prod}) and Water footprinting of consumption sector (WF_{cons}). The WF_{prod} refers to the amount of water utilized for the production of goods and services exported to foreign countries and the WF_{cons} is the amount of water consumed for the production of goods and services which is consumed by the local people. The water footprints are basically distinguished into blue, green and grey footprint components (Vanham and Bidoglio 2014). The water footprinting of the agricultural products quantified in the year 1996–2005 by Hoekstra and his team members in a spatially-explicit way revealed the following conclusions. The water footprint assessment is performed in the 5 by 5 arc minute grid using the CROPWAT model for nearly 200 food crops, fibres, flours, etc.

11.1 Blue Water Footprint Calculation

The calculation of the blue water footprint is the primary step in assessment of the water footprint of the agricultural products. This blue water footprint is basically indicated as a marker of wasteful utilization of alleged blue water such as surface water or ground water. Usually, the water dissipated and water not returning back to the catchment area within a particular period of time falls under the immoderate water utilization category. The improper usage of the water sources leads to the water scarcity so the monitoring of the utilization of water is a very essential step in the water footprint assessment. Water in streams and aquifers can be utilized for water system or mechanical or residential purposes. Be that as it may, in a specific period one can't devour more water than is accessible. The blue water footprint measures the amount of water accessible in a specific period that is expended in the agricultural activity. Water is a sustainable asset; however that does not imply that its accessibility is boundless. The blue water footprint is measured in terms of quantity of water consumed per unit of product. The blue water footprint is calculated using the formula:

$$WF_{blue} = \text{Blue water evaporation} + \text{Blue water incorporation} + \text{Lost return flow}$$

11.2 *Green Water Footprint Calculation*

The green water footprint is a pointer of the water utilization by human beings other than the surface and ground water. Green water alludes to the precipitation ashore that does not keep running off or revives the groundwater but rather is put away in the dirt or incidentally remains over the dirt or vegetation. In the end, this piece of precipitation vanishes or happens through plants. Green water footprint is basically affected by the environmental issues such as global warming and emission of greenhouse gases. This, negative impacts of environment leads to the modification in climatic pattern of the earth. This measurement of green water is more essential in the cultivation of crops and plants. The green water footprint is the volume of water devoured amid the manufacturing procedure. The green water footprint is calculated using the formula such as:

$$WF_{green} = \text{Green water evaporation} + \text{Green water incorporation}$$

In the calculation of green water footprinting, two main components are essential such as amount of green water evaporated and the green water incorporated. The former component is the quantity of water evaporated from the run-off water which is collected from the rain water, reservoirs, lakes, etc.; the other component is the measure of the water sources incorporated in the production of various items. The green water footprint also takes into account about the contaminations which are causing alterations in the quality and quantity of water. Green water utilization in agriculture can be estimated or assessed with an arrangement of exact equations which are calculating the evapotranspiration in view of information on atmosphere, soil and product.

11.3 Grey Water Footprint Calculation

The grey water footprint of a procedure step is a marker of the level of freshwater contamination that can be related with the manufacturing of products. It is characterized as the volume of freshwater that is required to absorb the heap of toxins in light of existing encompassing water quality benchmarks. It is figured as the volume of water that is required to weaken toxins to such a degree, to the point that the nature of the surrounding water stays above concurred water quality benchmarks. The grey water is also known as the dilution water which is used to dilute the heavy metals and toxins released in the effluents after all the manufacturing process. Decreasing the chances of eliminating the heavy toxin waste in the industries will surely decrease the release of unwanted toxins and chemicals into the water sources. The grey water footprint is used as a marker of contamination in the water sources. The grey water footprint is measured using the formula such as:

$$WF_{grey} = \frac{L}{C_{max} - C_{nat}}$$

The L is the concentration of the pollutant mixed in the water sources and C_{max} is the concentration of the pollutants that are permissible in the water and C_{nat} is the concentration of natural contaminants in the receiving body. The measurement of the pollutants in the receiving water is done in both direct and indirect manner. Not all the contaminants are directly released in the industrial effluents; some of the contaminants are released through various indirect ways. At the point when chemicals are straightforwardly discharged into a surface water body, the heap can specifically be estimated. The common focus in a getting water body is the fixation in the water body that would happen if there were no human unsettling influence in the catchment (Hoekstra et al. 2009).

12 Water Footprinting of Various Crops Produced

The normal water impression per ton of essential yield contrasts fundamentally among crops and crosswise over creation districts. Harvests with a high return or extensive portion of product biomass that is collected for the most part have a littler water impression for each ton than crops with a low yield or little division of harvest biomass collected.

According to the results published by Mekonen and Hoekstra (2011), The water footprinting of various primary food crops are listed below:

The water footprinting of vegetable oil are:

- Sunflower oil—6800 m^3 ton^{-1}
- Cotton seed oil—3800 m^3 ton^{-1}
- Castor oil—24700 m^3 ton^{-1}
- Ground nut oil—7500 m^3 ton^{-1}
- Maize oil—2600 m^3 ton^{-1}

The water footprinting of some of the food crops are:

- Wheat—1827 m^3 ton^{-1}
- Maize—m^3 ton^{-1}
- Cereals—1644 m^3 ton^{-1}

The water footprinting of some of the fruits are:

- Peach—910 m^3 ton^{-1}
- Watermelon—235 m^3 ton^{-1}
- Apple—820 m^3 ton^{-1}
- Orange—560 m^3 ton^{-1}

The water footprinting of the major food crops used universally by humans for consumption are listed below:

- Sugar crops—200 m^3 ton^{-1}
- Cereals—1600 m^3 ton^{-1}
- Oil crops—2400 m^3 ton^{-1}
- Tubers—400 m^3 ton^{-1}
- Pulses—4000 m^3 ton^{-1}
- Spices—7000 m^3 ton^{-1}
- Nuts—9000 m^3 ton^{-1}

At the nation level, the biggest aggregate water impressions were evaluated for India, China, the USA, Brazil, Russia and Indonesia. These six nations together record for about half of the worldwide aggregate water impression identified with trim creation. The biggest green water impressions are likewise found in these six nations: India, China, the USA, Russia, Brazil furthermore, Indonesia (Mekonen and Hoekstra 2011).

13 Standards and Guidelines

The standards and guidelines are helpful in conservation and management of fresh water sources. The certification and labelling of the agricultural products are helpful in developing the trust of the consumers on the characteristics and features of the item produced. The standards and guidelines are very important in the food and agri businesses. The objective of the standards is to give a clear cut idea about the quality of the product produced. This standards and guidelines may be single dimensional or multi dimensional; depending on the nature of the product it takes into account every criteria of the product. Some of the standards and guidelines for water footprinting of agricultural products are explained as follows:

13.1 Water Footprint Network

The water footprint network is a standard organization for enunciating the standards for water footprint of various sectors and it was formulated by a group of participants in the year 2008–2011, Which remained as a great initiative for the final Global Water footprint standard in the year 2011 (Hoekstra 2017). Arjen Hoekstra is the first person to work with the water footprinting tool, while he was working in the IHE Institute of Water Education. He devised a tool to evaluate the volume of water utilized to produce various products and he also examined the volume of water contaminated due to the production of different goods and services. Some of the industries such as Heineken, SABMiller, Nestle, Pepsico, Coca-cola, Unilever have implemented some water conservation measures understanding their water reliance and the water-related hazard confronting their organizations. The Water Footprint Network was established by Hoekstra, in 2008 at University of Twente. This Water Footprint standard is established worldwide with the point of uniting a portion of the brightest personalities focused on showing how Water Footprint Assessment can enable us to defeat the difficulties of unsustainable water utilize and it is used by all business sectors, common society, multilateral society and scholarly associations (Water footprint network).

13.2 ISO 14046

ISO 14046 determines standards, prerequisites and rules identified with water footprinting assessment of items, procedures and associations in view of life cycle evaluation (LCA). It is a standard for evaluating the water sources and helpful in conserving the water sources. The ISO 14046 provides the many benefits in the water conservation and management. It allows the water footprint as a standard tool for studying the factors which have negative impact on the water sources. The ISO 14046 provides the following benefits:

- Assess and get ready for the future dangers to your water utilize
- Identify approaches to decrease the ecological effects of your water utilize
- Improve effectiveness at item, process and hierarchical levels
- Share information and best practice with industry and government
- Meet client desires of expanded natural obligation

Just air and soil outflows that effect water quality are incorporated into the appraisal, and not all air and soil discharges are incorporated. The consequence of a water impression evaluation is a solitary esteem or a profile of effect marker comes about. Though revealing is inside the extent of ISO 14046, correspondence of water impression comes about, for instance as names or announcements, is outside the extent of ISO 14046 (ISO 14046: 2014).

13.3 FAO/IEAE Programme

IAEA stands for International Atomic Energy Agency which is an intermediary organization in Vienna working on the nuclear energy, technology and science. It is established in the year 1957 and it supports the public by ensuring with intact and assured nuclear technologies. Around 70% of worldwide freshwater utilization is utilized as a part of the crop cultivation segment, yet water utilize productivity in numerous nations is underneath 50%. The FAO announces that before 2050, worldwide water necessities for agribusiness will elevates by 50 percent to meet the expanded sustenance requests of a developing populace. Atomic and isotopic strategies give information on the water consumption and water loss through soil dissipation and help upgrade water system planning and enhance water utilize effectiveness. Tending to these issues requires an enhanced administration of land and water. Mutually with the FAO, the IAEA helps Member States create and receive atomic based advances to enhance rural water administration hones that help the strengthening of yield generation and the protection of characteristic assets (IAEA).

14 Global Water Tool

Global water tool (GWT) was enunciated by WBCSD in the year 2007 and it is used to help all types of business activities in managing the consumption of water resources and to lower the water related risk. The GWT permits the industries to monitor their own water use with other factors such as quality, availability and other environmental conditions. This tool infers about the local water consumption and universal water consumption. This tool is available for free of cost and it helps the industries to monitor the location with high water level and water scarcity regions.

15 Progress and Challenges in the Water Footprinting of Agricultural Products

The ecological footprint and the water footprint are the two effective and famous tools for evaluating the natural sources, they both are similar in function but their systems are different. The ecological footprint is the volume of land region required to manage the population growth and it specifically characterizes the natural effect of human utilization by looking at the accessible bioproductive zone to the region required for the utilization of particular merchandise and ventures. The water footprint is the volume of freshwater required for the production of various items and products. The water impression tracks the volume of virtual water utilized by a populace, where virtual water is the volume of freshwater used to create a ware, decent or benefit along the different strides of production (Gleeson et al. 2012). The water footprinting in agricultural sectors is measured by considering of all the international, national, local sectors. Some of the measures can be followed to reduce the water wastage during the production of agricultural products. The water foot printing can be lesson by using a proper monitoring system for evaluating the consumption of water sources. Improving the worldwide crop cultivation approaches can help out during the dry periods. The water footprinting and water shortage information should be incorporated in an organized system and the benchmarks should be indicated as a precondition for venture. The type of crop cultivated and the condition at which the crop cultivation should be carried out must be planned before hand to decrease the water wastage during agriculture. Several new crops varieties should be cultivated based on the appropriate climatic conditions with a specific goal to diminish water system requirement. Decreasing the utilization of manures, pesticides and bus sprays can lessen the chemical contamination in the water framework. The afforestation and reforestation activities increase the water requirements in that region (UNEP 2011). The studies related to the consumption and dissipation of fresh water sources has increased everyday and the interest for water conservation has increased nowadays. Initially the water utilization for agriculture is calculated without the water foot printing tool and now it has given immense importance for evaluating the global water demand in agricultural sector (Lovarelli et al. 2016). Initially, the water footprinting of wheat is examined based on the three various spatial resolution levels such as national, provincial and grid level. The grid level is selected as the effective resolution level for determining the water footprint of wheat (Mekonnen and Hoekstra 2014). Water speaks to a principle asset for human exercises and its administration. Together with sustenance and vitality creation, water shortage is a standout amongst the most extraordinary issues for society. In most European locales rural preparations are concentrated water buyers. Hence, these frameworks are extremely powerless against a potential diminishing in water accessibility. In 2011 a European Concerted Research Action (COST) began, went for the Assessment of EUROpean AGRIculture WATer utilize and exchange under environmental change (EURO-AGRIWAT). EURO-AGRIWAT expects to plan and scatter suggestions and rules for empowering a more proficient water asset administration in connection with horticultural exercises under environmental change and inconstancy, and includes coordinated effort amongst researchers and partners (Water footprinting network). The water footprinting of the agricultural products varies based on the type of crop cultivated. Harvests with a high return or vast division of product biomass that is gathered for the most part have a littler water footprinting for every ton than crops with a low yield or little part of harvest biomass gathered. The agricultural products such as Tea, Coffee, spices, nuts, fibres tobacco and rubber possess a large water footprint (Mekonnen and Hoekstra 2011).

16 Future Trends

The water conservation and management is important to save some water sources for the future generation. Advancement in technology has led to the development of many new strategies for lowering the water footprint. Many alternative strategies have been discovered to lower the pollution level and the water scarcity. Implementing stringent rules and regulations by government can help in decreasing the problems caused to the water sources. In the future, around 2025, it is estimated that nearly 67% of the total world's population will suffer due to water scarcity. Due to several environmental issues such as global warming and emission of green house gases, the climatic pattern may vary which automatically leads to the water scarcity and pollution. As the urbanization and industrialization gets advanced the probabilities of water availability gets lowered. Due to deficiency of rainfall, the agriculture will get affected and the yield will get decreased. The decrease in the food production rate will directly lead to poverty. Water footprinting is implemented in every sector to monitor the level of fresh water sources in the earth.

17 Conclusion

Water is the major source for all the human activities and agriculture is the backbone of food production. Agriculture provides the one of the basic need of the human which is called as food and it nourishes living beings with essential nutrients. Water footprinting of agricultural products is necessary to monitor the negative impacts of environment on the water sources. The green, blue and grey water footprints segregate the reasons for the decrease in water availability. Around 60% of the total world's water is contributed in agriculture and it is not renewable. The contaminants which affect the quality and quantity of the fresh water sources should be monitored using the water footprint. The Four phases of water footprint helps in understanding every criteria affecting the water sources. Implementation of new standards and policies can decreases the wastage of water and can help in fighting against the water scarcity.

References

Agricultural Water Management. Joint FAO/ IAEA programme. https://www.iaea.org/topics/agricultural-water-management.

Assessment of the efficiency of the water footprinting approach and of the agricultural products and foodstuff labelling and certification schemes. (2011). European commission-Direct general environment, RPA. http://ec.europa.eu/environment/water/quantity/pdf/Final%20Report%20Part%20A%20%20clean%2027%20Sept_2011%202.pdf.

Assessment of the Efficiency of the Water Footprinting Approach and of the Agricultural Products and Foodstuff Labelling and Certification Schemes. (2011). European Commission Directorate-General Environment, September 2011.

Centers for Disease Control and Prevention. (2006). Other uses and types of water in 2006. https://www.cdc.gov/healthywater/other/agricultural/index.html.
Chukalla, A. D., Krol, M. S., & Hoekstra, A. Y. (2018). Trade-off between blue and grey water footprint of crop production at different nitrogen application rates under various field management practices. *Science of the Total Environment, 626,* 962–970.
Gleeson, T., Wada, Y., Bierkens, M. F. P., & Beek, L. P. H. V. (2012). Water balance of global aquifers revealed by groundwater footprint. *Nature, 488,* 197–200.
Herath, I., Green, S., Singh, R., Horne, D., Zijpp, S. V. D., & Clothier, B. (2013). Water footprinting of agricultural products: A hydrological assessment for the water footprint of New Zealand's wines. *Journal of Cleaner Production, 41,* 232–243.
Hoekstra A. Y. (2003). *Virtual water trade: Proceedings of the International Expert Meeting on Virtual Water*, Value of Water Research Report Series No. 12.
Hoekstra, A. Y. (2017). Water footprint assessment: Evolvement of a new research field. *Water Resource Management, 31,* 3061–3081.
Hoekstra. A. Y., Chapagain. A. K., Aldaya. M. M., & Mekonnen. M. M. (2009). Water footprint manual, state of art 2009, Water footprint report, November 2009.
Hoekstra, A. Y., Chapagain, A. K., Aldaya, M. M., & Mekonnen, M. M. (2011). *The water footprint assessment manual: Setting the global standard*. London: Earth scan.
Hogeboom, R. J., Knook, L., & Hoekstra, A. Y. (2018). The blue water footprint of the world's artificial reservoirs for hydroelectricity, irrigation, residential and industrial water supply, flood protection, fishing and recreation. *Advances in Water Resources, 113,* 285–294.
ISO 14046—Water footprint principles, requirements and oppurtunities. https://www.bsigroup.com/en-IN/ISO-14046-Water-footprint/.
Juan, J. A. D., Tarjuelo, J. M., Ortega, J. F., Valiente, M., & Carrioan, P. (1999). Management of water consumption in agriculture a model for the economic optimisation of water use: Application to a sub-humid area. *Agricultural Water Management, 40,* 303–313.
Lovarelli, D., Bacenetti, J., & Fiala, M. (2016). Water footprint of crop productions: A review. *Science of the Total Environment, 548–549,* 236–251.
Mekonnen, M. M., & Hoekstra, A. Y. (2011). The green, blue and grey water footprint of crops and derived cropproducts. *Hydrology and Earth System Sciences, 15,* 1577–1600.
Mekonnen, M. M., & Hoekstra, A. Y. (2012). A global assessment of the water footprint of farm animal products. *Ecosystems, 15,* 401–415.
Mekonnen. M. M., & Hoekstra. A. Y. (2013). Water footprint benchmarks for crop production, UNESCO-IHE, December 2013.
Mekonnen, M. M., & Hoekstra, A. Y. (2014). Water footprint benchmarks for crop production: A first global assessment. *Ecological Indicators, 46,* 214–223.
Vanham. D., & Bidoglio. G. (2014). The water footprint of agricultural products in European river basins, *Environmental Research Letters*, *9*.
Water Footprint and Corporate Water Accounting for Resource Efficiency. (2011). UNEP.
Water Use in Agriculture. (2016). OECD Council Recommendation on Water. http://www.oecd.org/agriculture/water-use-in-agriculture.htm.
Water Footprinting Network. http://waterfootprint.org/en/water-footprint/what-is-water-footprint/.

Water Footprint of Livestock Farming

I. Noya, L. Lijo, O. Piñeiro, R. Lopez-Carracelas, B. Omil, M. T. Barral, A. Merino, G. Feijoo and M. T. Moreira

Abstract In recent times, society has been confronted with problems related to food safety and environmental protection. As a result, consumers are increasingly demanding an evolution towards the consumption of healthy products, produced within a framework of protection of natural resources and the environment. As a primary source of protein and micronutrients, meat products are considered essential elements of the human diet. In the search for products that meet the criteria of excellence of natural products produced in a sustainable way, the Community of Natural Parks of Galicia (Northwest Spain) is a territory recognized and protected as a region of magnificent landscapes and natural ecosystems. In this context, a framework for action has been planned to define environmental sustainability criteria for the award of a specific eco-label for products produced in these ecosystems. In particular, the sustainable use and economic development of the territory through livestock farming is proposed. To this aim, an environmental analysis has been carried out to identify and compare livestock systems based on extensive practices for beef production in the framework of the Galician Natural Parks. To this end, several environmental indicators were selected, although special attention was paid to two of them because of their particular relevance in assessing the environmental profile of livestock products: Carbon Footprint (CF) and Water Footprint (WF). The principles of the Life Cycle Assessment (LCA) were applied by the former, while the Water Footprint Network (WFN) guidelines were followed in the latter case. According to the results obtained for the different indicators, feed production (including grass, cereals and concentrated feed) contributed significantly to the overall impacts of the supply chain (up to the farm gate), regardless the environmental indicator evaluated. Considering the

I. Noya (✉) · L. Lijo · G. Feijoo · M. T. Moreira
Department of Chemical Engineering, University of Santiago de Compostela,
15782 Santiago de Compostela, Spain
e-mail: isabel.noya@usc.es

O. Piñeiro · B. Omil · M. T. Barral · A. Merino
Department of Agricultural Chemistry, University of Santiago de Compostela, Lugo, Spain

R. Lopez-Carracelas
Dirección Xeral de Patrimonio Natural, Consellería de Medio Ambiente e Ordenación do Territorio, Xunta de Galicia, Santiago de Compostela, Spain

S. S. Muthu (ed.), *Environmental Water Footprints*, Environmental Footprints and Eco-design of Products and Processes,
https://doi.org/10.1007/978-981-13-2508-3_2

key role played by the production of the various ingredients for the formulation of feeds and their associated transport, the need to promote more sustainable production methods, as well as the exploitation of agricultural land adjacent to the farms, should be stressed, so that the environmental impacts are even lower than those estimated in this study. In addition, diffuse emissions also had a significant impact on most LCA-based categories, mainly due to the emission of nitrogen compounds to air from manure storage and their application to agricultural soils. Similarly, water demand from both irrigation steps in cultivation practices and on-farm activities had a critical role in WF results. From these outcomes, it would be desirable for waste management (especially livestock manure) to be carried out in accordance with principles of minimum impact, valuing the different streams to obtain reclaimed water for irrigation and bio-fertilizers that can replace those of chemical synthesis. Finally, the main results were compared with published works (in terms of CF and WF ratios) focusing on beef products and lower environmental impacts were registered. On this basis, common criteria for eco-labelling in beef production systems were defined in the framework of the Galician natural areas.

Keywords Beef production · Carbon footprint · Water use · Environmental profile Eco-label · Natural parks · Galicia

Abbreviations and Acronyms

ALO	Agricultural Land Occupation
CF	Carbon Footprint
EEA	European Ecological Label
FAO	Food and Agricultural Organization
FD	Fossil Depletion
FE	Freshwater Eutrophication
FET	Freshwater Ecotoxicity
FU	Functional Unit
GHG	Greenhouse Gas
HT	Human Toxicity
IPCC	Intergovernmental Panel on Climate Change
IR	Iionising Radiation
ISO	International Organization of Standardization
LCA	Life Cycle Assessment
LCI	Life Cycle Inventory
ME	Marine Eutrophication
MET	Marine Ecotoxicity
MRD	Mineral Resource Depletion
NI	Normalized Index
NLT	Nataul Land Transformation
NW	Northwest

OD	Ozone Depletion
OECD	Organization for Economic Cooperation and Development
PMF	Particulate Matter Formation
POF	Photochemical Oxidant Formation
TA	Terrestrial Acidification
TET	Terrestrial Ecotoxicity
ULO	Urban Land Occupation
WD	Water Depletion
WF	Water Footprint
WFN	Water Footprint Network

1 Introduction

Livestock farming is one of the sectors with the greatest economic and social impact on food production, where various socioeconomic dynamics have been developed, which have led to models of organization, management and marketing of meat (Alexandratos and Bruinsma 2012; de Vries and de Boer 2010; Thornton 2010). Meat consumption is related to living standards, diet, livestock production and consumer prices, as well as to macroeconomic uncertainty (OECD/FAO 2017). The demand for meat is associated with higher incomes and a shift—due to urbanization—in food consumption that favour the increase of animal proteins in diets (OECD/FAO 2017). So much so that in the last forty years, its demand has increased significantly worldwide, reaching a ratio of around 63% within European borders (Ciolos 2012; FAO 2014). Furthermore, among the main varieties in Europe, beef is one of the most widely consumed (11.1 kg of beef per capita in 2016), as reported in OECD-FAO Agricultural Outlook in 2017 (OECD/FAO 2017). These values are high, but relatively far from the ranking of the countries with the highest consumption of beef (beyond European boundaries): Uruguay, Argentina and Hong Kong; these three countries consumed more than 50 kg of beef per capita (Beef Market Central 2016). However, in contrast to its socio-economic relevance, beef production is also responsible for several environmental consequences, especially in terms of climate change and land use change (de Vries et al. 2015; Gerber et al. 2015). In fact, it accounts for about 41% of total emissions from the livestock sector in Europe, with greenhouse gas (GHG) emissions ranging from 14 to 32 kg CO_2 equivalents delivered to air (de Vries et al. 2015; Gerber et al. 2013; Opio et al. 2013; Steinfeld et al. 2006). On the other hand, beef production is also one of the dominant in recent land degradation and deforestation (Cederberg et al. 2011). This highlights the need for an in-depth study of the environmental burdens associated with the beef production chain in order to move towards greater economic and environmental competitiveness and sustainability (Notarnicola et al. 2012).

In this context, the Spanish livestock sector has experienced a surprising and positive development in recent years (INTERAL 2008; MAPAMA 2017). Indeed, livestock farming contributes 40% to Spanish agricultural production (Rodríguez

Casado et al. 2009). Focusing on meat production, Spain has developed an intensive and highly efficient meat sector in recent decades, which has led to enormous growth in pork and, to a lesser extent, poultry and calves, to the detriment of extensive livestock farming (INTERAL 2008). This is reflected in the predominance of the use of compound feed, with an annual production of more than 20 million tonnes and around 14,000 workers (MAPAMA 2017). Consideration of the link between feed production and livestock farming is therefore essential from both economic and environmental perspectives (Galloway et al. 2007).

Regarding beef production, it has historically played a critical role within the Spanish national livestock farming, closely linked to the social fabric of rural areas (Rodríguez Pascual 2008). In fact, Spain has been very competitive in beef production since it joined the EEC in 1986 (INTERAL 2008). Currently, Spain has undergone an upward trend in livestock production that dates back to 2014, while experts predict that it will continue to grow in the coming years (Valverde 2017). This situation can mainly be attributed to increased demand from non-EU countries, together with the stability achieved in the prices of feed materials (Valverde 2017). In this context, beef production in Spain increased by 1.8% in 2016 compared to the previous year, representing more than 2 million head of cattle slaughtered in the entire national agricultural area (Valverde 2017). It highlights the greater importance that young cattle (animals under 12 months) and suckler cows have acquired in the country, to the detriment of dairy cows (Rodríguez Casado et al. 2009).

Focusing on the environmental framework, numerous studies can be found in the literature that show that dietary intake of meat proteins can account for most of their ecological footprint, water pollution and scarcity as well as carbon dioxide (CO_2) emissions (Rodríguez Casado et al. 2009; Roy et al. 2009). Although the conversion rate between kilograms of feed consumed and kilograms of meat produced has improved substantially in recent decades, livestock farming is still responsible for 9% of CO_2 emissions from human activities and produces 65% of nitrogen oxides (NO_x) emissions, mainly associated with manure management (FAO 2006; Rodríguez Casado et al. 2009). Therefore, the increase in meat production implies, as a general rule, several environmental problems, mainly associated with the pollution of water environmental and/or GHG emitted into the atmosphere. Within this framework, improvement initiatives have also taken the lead in the search for more sustainable production schemes (Notarnicola et al. 2012).

In this regard, several environmental methodologies have recently been applied to assess the extent to which such initiatives could improve the environmental profile of conventional beef production practices (Aivazidou et al. 2016; de Vries et al. 2015). Most of them follow the life-cycle approach to assess common environmental indicators such as carbon footprint (CF), acidification and eutrophication potential (de Vries et al. 2015). In this regard, diffuse emissions from enteric fermentation and manure management were found as the largest contributors to CF results in most cases (Beauchemin et al. 2010; Lupo et al. 2013; Mogensen et al. 2015), while manure emissions were found to have also a relevant influence in both eutrophication and acidification impacts (Lupo et al. 2013). However, lower GHG emissions were estimated in concentrate-based systems per unit of beef produced, in comparison with roughage-based alternatives, although no evidence was obtained with regard to eutrophication and acidification potentials (de Vries et al. 2015). These results reveal that mitigation strategies to reduce the related impacts of beef production systems should be directly linked to the manipulation of animal feed and the valorization of animal manure in agricultural activities (Lupo et al. 2013; Pelletier et al. 2010). Moreover, in the same vein, carbon sequestration of crops can also have a favourable effect on the environmental profile of beef production systems (Beauchemin et al. 2010; Mogensen et al. 2015; Pelletier et al. 2010); however, on the other hand, permanent use of arable land for feeding purposes can contribute to carbon changes and soil degradation (Mogensen et al. 2015). Therefore, there is also a need to improve land management practices to prevent ongoing land expansion and potential environmental damage (Cederberg et al. 2009; de Vries et al. 2015).

However, to date, much more limited information is available on beef production and its potential impact on water resources (de Vriest et al. 2015; Harding et al. 2017; Mekonnen and Hoekstra 2012; Ridoutt et al. 2012, 2014). While earlier publications on water use in the livestock sector can be found in literature (Chapagain and Hoekstra 2004; Hoekstra and Chapagain 2007; Molden et al. 2007; Pimentel et al. 2004; Renault and Wallender 2000; van Breugel et al. 2010), updated research works in the field are much fewer nowadays, also involving the beef sector (Peters et al. 2010; Ridoutt et al. 2014). This situation is less promising today at the Spanish level: although the water footprint (WF) of beef production has become a topic of common interest in European research, no studies have yet been carried out in Spain on this subject (Mekonnen and Hoekstra 2012). Accordingly, and also in line with the current trend in Europe, greater attention should be paid to the WF evaluation in pursuit of products that meet criteria related to food safety and environmental protection.

In this context, the Community of Natural Parks in Galicia (Northwest Spain), which represents protected areas of special environmental interest, aims to be a pioneer in obtaining a quality mark for extensive beef products. The sustainable use of vegetation by livestock farming has contributed to the maintenance of a high level of biodiversity and landscape quality, the prevention of forest fires and the socio-economic development of these rural areas. Due to the inherent characteristics of the National Parks, actions are carried out to improve pastures, livestock practices or fences, in order to maintain livestock that in turn contributes to the preservation of the natural environment. The commitment of the local action groups in the economic and social development of the area tries to prevent the abandonment of these areas.

The main objective of this study was to evaluate the environmental performance of the different beef production systems of the National Parks of Northwest Spain, paying special attention to CF and WF results. In addition, this work includes specific objectives such as promoting the dissemination of good farming practices and aims to raise awareness among the groups involved in the potentialities of sustainable livestock farming as a natural and economic resource; to energize groups on the economic possibilities offered by cooperation and access to quality labels; and to improve natural habitats dependent on livestock activity in order to ensure the conservation of species and achieve effective management of natural areas.

To this purpose, the first sections of the chapter defined the boundaries of the different target systems, together with the most relevant methodological options in environmental assessment. Special attention was then paid to the collection and processing of the inventory data, given their relevance to the reliability of the final results. Once the mass and energy balances had been defined, the WF values and other complementary environmental indicators were estimated and evaluated in the results sections. Finally, the robustness of the main outcomes was checked by comparison with similar studies in the literature.

2 Materials and Methods

The principles of the Life Cycle Assessment (LCA) philosophy (ISO 140402006; ISO 140442006; ISO 14067 2013) linked to the Water Footprint Network (WFN) guidelines (Hoekstra et al. 2011) were followed in this study for LCA and WF assessment, respectively.

2.1 *Goal and Scope Definition*

As aforementioned, the study focused on deepening the environmental performance of beef production systems according to extensive farming practices in the framework of the National Parks of NW Spain (Galicia). In this regard, a total of six natural areas were defined in this region (Fig. 1), in which a total of 12 municipalities participate distributed throughout the Galician territory (Table 1).

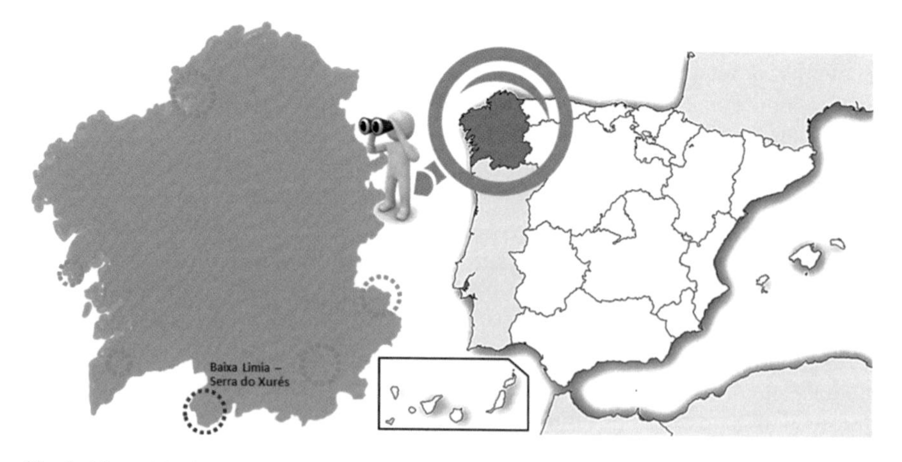

Fig. 1 Natural Parks in Galicia (NW Spain). Location of target systems: Baixa Limia—Serra do Xurés

Table 1 Main characteristics of Galician Natural Parks: denomination, area, number of identified producers and primary products generated (Xunta de Galicia 2018)

Galician natural parks	Area (ha)	Identified producers	Main primary products
Fragas do Eume	9,126	9	Honey and turnip green
Baixa Limia—Serra do Xurés	29,762	8	Honey, oil and beef meat
Serra da Enciña da Lastra	3,152	2	Honey, wine and chestnut
Complexo Dunar de Corrubedo e Lagoas de Carregal e Vixen	996	1	Honey
Monte Aloia	746	0	–
O Invernadeiro	5,722	0	–

While some parks are an important opportunity for the production of alternative primary products (mainly honey), local generation of market products has not yet been identified in others to date. As indicated in Table 1, beef production is specifically located in the Baixa Limia—Serra do Xurés Natural Park, the largest (by far) with a significant diversity of products, compared to the rest of the Galician parks. It is characterised by the intensive forest use of the soil, together with a significant increase in the delimitation of permanent grasslands directly related to extensive livestock practices.

Regarding the scope of the study, a cradle-to-farm gate perspective was followed, taking into account all the life cycle stages of the production chain until the cattle are ready to leave the farm for the slaughterhouse premises. Consequently, the subsequent stages of slaughter, processing, distribution, consumption and end-of-life were left outside the system boundaries. However, previous related studies in the literature have highlighted the reduced influence of such post-farm stages in relation to the impact of the entire productive chain (Daneshi et al. 2014; Fantin et al. 2012; González-García et al. 2013; Vasilaki et al. 2016).

2.1.1 Functional Unit (FU)

The FU was chosen in this study according to the main objective of the systems evaluated: cows breeding for beef production. With this in mind, "1 kg of beef carcass at farm gate" was defined as the basis for estimation and comparison. This choice would also be in line with other similar studies on livestock systems published to date, in which mass-based FUs prevail over other alternatives (de Vries et al. 2015; Gerber et al. 2015; McAuliffe et al. 2016).

2.1.2 System Description

Five different systems were evaluated in the study (Table 1); however, regardless of their particularities, all systems share the same scheme and structure. Accordingly, as shown in Fig. 2, each beef production system (foreground system) can be divided into two main subsystems: animal feed production (SS1) and farm management (SS2). Background processes were also taken into consideration (background system) as primary source of supply for the overall system, including crops cultivation, energy generation and chemicals manufacture.

The feed production subsystem (SS1) includes all processes related to the production of the different ingredients delivered to the cattle herd in each farming system (Table 2). Thus, according to the information provided by producers, both dried grass and chopped maize, whether produced on-farm or purchased, are supplied to cows (and bulls, where applicable), while concentrate feed (e.g. fodder) is specifically consumed by calves at farm. This fodder is mainly composed of maize, barley, rye and soybean in similar proportions.

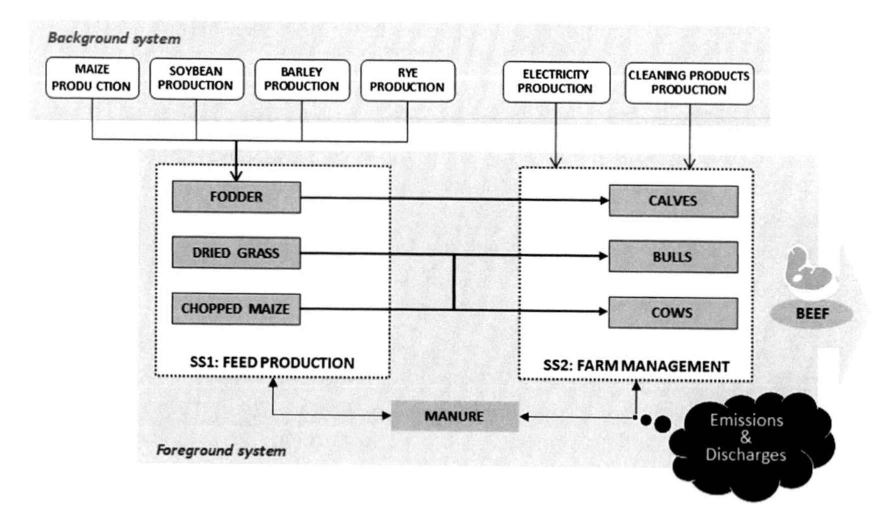

Fig. 2 General scheme of the processes included within the system boundaries for the five systems evaluated involving beef production

Table 2 Diets of cattle herd in the different farming systems (annual basis)

System	Dried grass (t/yr)	Chopped maize (t/yr)	Fodder (t/yr)
System 1 (S1)	120	30	40
System 2 (S2)	210	2	2
System 3 (S3)	10	40	3
System 4 (S4)	300	–	60
System 5 (S5)	75	–	48

Table 3 Agricultural activities involved in grass and maize cultivation (per ha) in the different systems

Cropping system	Month	Agricultural machinery			Input rates
		Weight (kg)	Effective work (ha/h)	Diesel use (kg)	
Dried grass					
Organic fertilization	N.D.	4,000	5	4	1 t manure
Mowing	N.D.				–
Mineral fertilization[a]	April				600 kg NPK
Loading	N.D.				–
Chopped maize					
Organic fertilization	April				10 t manure
Ploughing	May				–
Harrowing	May				–
Sowing	May	4,000	5	18	12 kg seeds
Weed control	June				2 L pesticide
Irrigation (2 steps)	August				9,000 L water
Chopping	August				–

[a]Only in case of S5;
N.D.: no data available

Grass production involves three main stages (Table 3): organic fertilization (about 1 t of cow manure per ha), mowing, mineral fertilization (only in one system) and final grass loading. Maize cultivation takes place over six months of the year (from April to August) and, compared to grass production, involves some additional agricultural activities: organic fertilization (10 t of cow manure per ha), ploughing, harrowing, sowing, weed control, irrigation and chopping. A tractor of 4,000 kg (100 kW of power) was assumed to be used in both cases, capable of working around 5 ha of agricultural soils per effective work-hour; however, diesel consumption varies between 5 and 18 kg per ha for grass and maize cultivation, respectively.

As far as SS2 is concerned, farm management comprises all processes related to cattle breeding at farm. The number of cows in each system varies from 50 to 122, and the number of calves from 40 to 90; 5 bulls can also be found in S5 (Table 4). This subsystem also includes land requirements for both animal feed production and farm facilities, as well as on-farm energy use. Finally, the manufacture of the different input flows, including drinking water, cleaning products and other agrochemicals, as well as the drugs used for cattle worming, were also considered in SS2, along with the related background processes (Fig. 2).

In contrast, the construction and maintenance of the infrastructure was not included within the system boundaries, as its contribution was expected to be negli-

Table 4 Main characteristics of each system under evaluation (annual basis)

System	Total area (ha)	Farm facilities (m^2)	Herd size (heads/yr)		Production yield (t beef/yr)	
			Cows	Calves	Live weight	Carcass weight
S1	120	800	120	85	25.2	14.0
S2	100	1,050	105	90	10.8	7.2
S3	80	900	50	40	14.4	8.0
S4	120	1,000	120	90	28.8	16.0
S5	106	2,000	127[a]	65	21.6	12.0

[a]5 bulls (122 cows) are included in S5

gible in line with other reports available in the literature on similar livestock systems (de Léis et al. 2015; Hoekstra et al. 2011; Lupo et al. 2013; Ridoutt et al. 2014).

2.1.3 Allocation Rules

Allocation can be defined as the assignment of the environmental impacts of a global system among its different functional products (ISO 14040 2006; Suh et al. 2010). However, in accordance with ISO standards, it is recommended that allocation be avoided, with priority given to dividing the system into subsystems or extending the system boundaries to incorporate greater functionality of potential co-products (ISO 14040 2006).

In this regard, allocation was not necessary in this study because beef (carcass weight) at farm gate was assumed to be the only product responsible for the impacts of the entire system. Cow manure is also generated as a result of livestock breeding at farm; however, it was not considered as a product of the system, since it has no economic value to local producers, but is required in on-farm cultivation (dried grass and chopped maize—Table 3) as an organic fertilizer. It was therefore estimated that related on-field nitrogen emissions (N_2O, NH_3, NO_x, NO_3^-) due to manure management and further application to soils would be included in the environmental assessment (see Table 5).

2.2 *Life Cycle Inventory (LCI) Analysis*

As explained above, the different systems under study include all the processes necessary for the production of the FU (1 kg of carcass beef at farm gate), as well as the manufacture of all the necessary inputs and the final management of the waste generated.

Table 5 LCI data (per FU) for each system under evaluation involving beef production

Inputs/Outputs	S1	S2	S3	S4	S5	Units
Inputs from environment						
Total area	85.7	139	100	75.0	88.3	m^2
Farm facilities	0.06	0.15	0.11	0.06	0.17	m^2
Water (from a water well)	267	494	205	240	292	L
Inputs from technosphere						
Materials						
Fodder (calves)	2.86	0.28	0.38	3.75	4.00	kg
Dried grass	8.57	29.2	1.25	18.8	6.25	kg
Chopped maize	2.14	0.28	5.00	–	–	kg
Antiparasitic treatment	0.17	0.29	0.13	0.15	0.21	mL
Bleach (solution)	–	–	–	–	0.13	mL
NPK	–	–	–	–	50.0	g
Energy						
Electricity	–	–	1.22	0.61	0.73	kWh
Outputs to technosphere						
Products						
Beef (carcass weight)	1.00	1.00	1.00	1.00	1.00	kg
Beef (live weight)	1.80	1.50	1.80	1.80	1.80	kg
Waste						
Cow manure[a]	12.9	25.0	12.5	11.3	15.0	kg
Outputs to environment						
Air emissions						
CH_4	549	933	400	480	677	g
N_2O	1.13	2.19	1.10	0.99	1.43	g
NH_3	15.6	30.4	15.3	13.7	19.2	g
NO_x	3.07	5.97	2.99	2.67	2.81	g
Water emissions						
NO_3^-	95.1	185	92.5	83.3	137	g

[a]Cow manure properties (assumed): nitrogen content = 5.57 g N/kg manure; dry matter = 15%

Table 6 Data sources used in LCA results

Inputs/Outputs	Data sources	Detailed information
Land requirements	Primary data[a]	–
Water	Primary data[a]	–
Fodder production	Primary data[a]	Ingredients and composition
	Ecoinvent® database	Background inventory: crops cultivation
Dried grass	Ecoinvent® database	Grass, organic {CH}—grass production
Chopped maize	Primary data[a]	Agricultural activities, inputs supply
	Ecoinvent® database	Background inventory: Althaus et al. (2007); Nemecek and Käggi (2007)
Antiparasitic	Ecoinvent® database	Chemical, inorganic {GLO}
		Tap water {RER}
Bleach (solution)	Ecoinvent® database	Sodium hypochlorite, without water, in 15% solution state {GLO}
		Water, deionised, from tap water, at user {GLO}
NPK	Adapted from ecoinvent® database	Nitrogen fertilizer, as N {GLO}
		Phosphate fertilizer, as P_2O_5 {GLO}
		Potassium fertilizer, as K_2O {GLO}
Electricity	Ecoinvent® database	Spanish country mix: Dones et al. (2007)
Cow manure	Primary data[a]	–
Carbon emissions	IPCC (2006)	Tier 2
Nitrogen emissions	IPCC (2006)	Tier 1/Tier 2

[a]Questionnaires fulfilled by farmers

Given the relevance of data quality in an LCA study, primary inventory data was always prioritised in the present study, minimizing the information collected from secondary sources. In this context, the primary data needed to construct the LCI were obtained through surveys distributed among the different producers. These surveys included data for the 2016/2017 agricultural year on beef production yield and type of holding, number of heads, animal feed, water consumption, use of chemicals, electricity consumption and waste production (mainly cow manure). The main inventory information (per FU) for the different systems is shown in Table 5; the main data sources are included in Table 6.

Additionally, it was necessary to estimate some data as it was not possible to obtain them by any other means. Firstly, methane (CH_4) emissions from the enteric fermentation of cows were calculated according to the guidelines provided by the Intergovernmental Panel on Climate Change (IPCC 2006). Thus, the Tier 2 method was used by combining default emission factors with the primary information recorded in the questionnaires.

In addition, emissions from manure application were also included according to the IPCC (2006) method. In particular, this source of nutrients for agricultural soils was considered to be responsible for significant air emissions in terms of dinitrogen monoxide (N_2O), ammonia (NH_3) and nitrogen oxides (NO_x), mainly derived from the nitrogen content of cow manure. Similarly, water emissions due to nitrates (NO_3^-) discharged to water environments from manure were also taken into account. However, for nitrogen-based emissions, both Tier 1 and Tier 2 methods were applied together on the basis of the availability of reliable primary information (IPCC 2006). In addition, it is important to note that the manure generated on each farm is not sufficient to meet the fertilization needs of the animal feed crop on the farm and therefore a surplus of manure is required. As a result, not only the emissions from manure management on the farm were included in the assessment, but also the additional organic fertilization required for crop growth, which also involves direct and indirect nitrogen emissions.

Finally, the ecoinvent® database (Althaus et al. 2007; Dones et al. 2007; Nemecek and Käggi 2007; Wernet et al. 2016) was also used to calculate the production of the different inputs used in each farming system. Thus, data on electricity generation (Spanish profile), as well as on the manufacture of chemical products used in cleaning and animal health care, together with emissions from diesel consumption have been extracted from this database.

As for the evaluation of WF, priority was again given to primary information. Thus, data on water consumed directly throughout the different stages of the life cycle of each system were collected from questionnaires filled in by beef producers (see Table 5). On the other hand, indirect water consumption was also quantified through the use of water from agricultural activities, feed production (including fodder), energy demand, the manufacture of chemicals and the disposal and management of final waste. However, in this case, secondary data sources were used mainly due to the absence of primary data. Accordingly, WF ratios for the cultivation of animal feeding (including crops for fodder production) were taken from Mekonnen and Hoekstra (2010); similarly, legal limitations published in the literature were considered to estimate the grey WF results. Thus, while the maximum acceptable (C_{max}) concentration values reported by Directive 91/271/EEC (1991) were taken into account in the calculations, the natural concentration (C_{nat}) in the receiving water body was assumed to be 0 mg/L. Finally, the ecoinvent® database was used to complete the background inventory, including the manufacture of chemicals and processes for energy generation. Secondary data sources in WF assessment are summarized in Table 7.

Table 7 Data sources used in WF results

Inventory data	Data sources
Cows feed	Mekonnen and Hoekstra (2010)
Dried grass	N.D.
Chopped maize	FAOSTAT crop code: 56
Calves feed (fodder)	Mekonnen and Hoekstra (2010)
Maize	FAOSTAT crop code: 56
Barley	FAOSTAT crop code: 44
Rye	FAOSTAT crop code: 71
Soybeans	FAOSTAT crop code: 236
Waste disposal	Franke et al. (2013)
Wastewater	Directive 91/271/EEC (1991)
Energy use	Ecoinvent® database (Dones et al. 2007)
Electricity	Spanish country mix—Electricity, low voltage {ES}
Chemicals	Ecoinvent® database (Althaus et al. 2007)
Antiparasitic	Chemical, inorganic {GLO}
	Tap water {RER}
Bleach	Sodium hypochlorite, without water, in 15% solution state {GLO}
	Water, deionised, from tap water, at user {GLO}

N.D.: no data available

2.3 Impact Assessment

SimaPro 8.2 software (PRé Consultants, Amersfoot, The Netherlands) was used for the computational implementation of the life cycle inventories (Goedkoop et al. 2013a). It is a professional tool that provides support for the calculation of the environmental impacts associated with a product or system from the LCA perspective (PRé Consultants 2016). In addition, this software contains several globally recognized databases (including ecoinvent®) along with the most relevant impact assessment methods, such as ReCiPe (Goedkoop et al. 2013b) and IPCC (2013), among others. Following experts' recommendations, the former represents the most updated alternative among similar assessment methodologies to date, while the IPCC guidelines provide a valuable common framework to evaluate the potential damages from GHG emissions and related global warming concerns (PRé Consultants 2016).

In this regard, the characterization factors provided by the ReCiPe Midpoint 1.12 (H) method were applied to estimate the life cycle environmental results in terms of terrestrial acidification (TA), freshwater eutrophication (FE) and marine eutrophication (ME) (Goedkoop et al. 2013b). Impacts from water depletion (WD) were also estimated and compared in relation to WF results. The three components of WF (green, blue and grey) were calculated and reported both individually and all together (Hoekstra et al. 2011). Moreover, particular attention was paid to climate

Table 8 Impact categories included in the NI results and related normalization factors according to the ReCiPe method (Goedkoop et al. 2013b)

Impact category	Acronym	Units	Normalized factor
Climate change	CC	kg CO_2 eq	$8.92 \cdot 10^{-5}$
Ozone depletion	OD	kg CFC-11 eq	45.4
Terrestrial acidification	TA	kg SO_2 eq	$2.91 \cdot 10^{-2}$
Freshwater eutrophication	FE	kg P eq	2.41
Marine eutrophication	ME	kg N eq	$9.88 \cdot 10^{-2}$
Human toxicity	HT	kg 1,4-DB eq	$1.59 \cdot 10^{-3}$
Photochemical oxidant formation	POF	kg NMVOC	$1.76 \cdot 10^{-2}$
Particulate matter formation	PMF	kg PM10 eq	$6.71 \cdot 10^{-2}$
Terrestrial ecotoxicity	TET	kg 1,4-DB eq	0.121
Freshwater ecotoxicity	FET	kg 1,4-DB eq	$9.09 \cdot 10^{-2}$
Marine ecotoxicity	MET	kg 1,4-DB eq	0.115
Ionising radiation	IR	kg U_{235} eq	$1.60 \cdot 10^{-4}$
Agricultural land occupation	ALO	m^2a	$2.21 \cdot 10^{-4}$
Urban land occupation	ULO	m^2a	$2.46 \cdot 10^{-3}$
Natural land transformation	NLT	m^2	6.19
Water depletion	WD	m^3	0.00
Mineral resource depletion	MRD	kg Fe eq	$1.40 \cdot 10^{-3}$
Fossil resource depletion	FD	kg oil eq	$6.43 \cdot 10^{-4}$

change concerns, so that CF results were estimated in accordance with IPCC factors (IPCC 2013). The choice of such environmental indicators was conducted in line with similar studies in the literature (McClelland et al. 2018). Moreover, the Normalized Index (NI) was also estimated as a basis for comparison among the different systems, since it provides a global view of their environmental performance. To this aim, all the categories of the ReCiPe method were evaluated as a whole, but taking into account the relative importance of each category according to its normalized factor (Table 8).

Thus, the value obtained for each category is relative to a reference value, giving rise to dimensionless categories; this reference value may refer to the results associated with such category at the global, national or regional level. Normalization factors provided by the ReCiPe Midpoint 1.12 (H) method were applied, involving European reference values. It should be noted that WD is attributed a zero standardization factor, as, unlike other resources, a global distribution involving water supply cannot be guaranteed. This means that water use may be responsible for significant impacts at the local level, but no models are available to estimate its damage in a broader context (Goedkoop et al. 2013a).

3 LCA Results

The life cycle impact assessment was carried out for each of the five systems for the different indicators under study. Normalized results and LCA impacts obtained by global and by sub-processes are displayed in Table 9 and Figs. 3, 4, 5, 6, 7 and 8. In this regard, these processes were grouped into the following main contributing factors to facilitate the interpretation of the environmental results: diffuse emissions, cow feed cultivation, fodder production, energy use, agrochemical production, land occupation and cleaning activities (only in case of S5). Additionally, water use was also taken into account in the analysis of the WD impact.

Diffuse emissions include both CH_4 emissions derived from enteric fermentation at farm and nitrogen-based emissions due to manure storage and its subsequent application to soils. Cow feed cultivation comprises all the impacts related to the cultivation of chopped maize and dried grass supplied to cows, while fodder production refers to the environmental burdens directly related to the manufacture of fodder consumed by calves. Energy use considers the impacts due to on-farm electricity requirements, agrochemical production comprises the manufacture of fertilizers and other chemicals consumed in agricultural activities, and land occupation reflects the impacts of land use. Finally, the use of cleaning products was considered in cleaning activities.

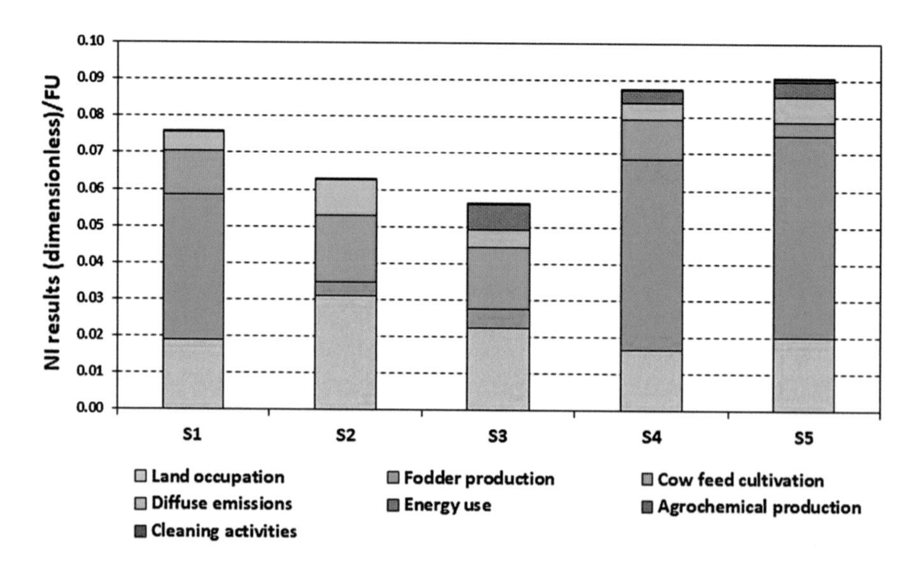

Fig. 3 NI results for the different processes involved in beef production in farming systems

Table 9 LCA and normalized results for the different systems per FU (1 kg of carcass beef at farm gate)

LCA results	S1	S2	S3	S4	S5	Units
CC (CF)—Total	*20.7*	*31.2*	*15.1*	*19.3*	*25.3*	*kg CO_2 eq*
Diffuse emissions	82.01	93.14	82.62	77.05	83.09	%
Cow feed cultivation	6.00	6.10	11.49	6.04	1.53	%
Fodder production	11.84	0.76	2.13	16.70	13.54	%
Energy use	0.00	0.00	3.77	1.46	1.35	%
Agrochemical production	0.00	0.00	0.00	0.00	0.49	%
Land occupation	0.15	0.00	0.00	0.20	0.00	%
Cleaning activities	–	–	–	–	0.00	%
TA—Total	*93.4*	*95.3*	*114*	*71.7*	*83.6*	*g SO_2 eq*
Diffuse emissions	42.85	81.72	34.31	48.96	58.69	%
Cow feed cultivation	34.82	16.15	60.21	10.43	2.99	%
Fodder production	22.34	2.13	2.41	38.18	34.91	%
Energy use	0.00	0.00	3.07	2.43	2.50	%
Agrochemical production	0.00	0.00	0.00	0.00	0.91	%
Land occupation	0.00	0.00	0.00	0.01	0.00	%
Cleaning activities	–	–	–	–	0.00	%
FE—Total	*0.94*	*0.51*	*0.43*	*1.36*	*1.31*	*g P eq*
Diffuse emissions	0.00	0.00	0.00	0.00	0.00	%
Cow feed cultivation	19.51	85.58	35.70	20.34	7.02	%
Fodder production	80.49	14.42	23.32	73.23	80.89	%
Energy use	0.00	0.00	40.98	6.43	7.99	%
Agrochemical production	0.00	0.00	0.00	0.00	4.09	%
Land occupation	0.00	0.00	0.00	0.00	0.00	%
Cleaning activities	–	–	–	–	0.00	%

(continued)

Table 9 (continued)

LCA results	S1	S2	S3	S4	S5	Units
ME—Total	*67.4*	*51.0*	*62.1*	*58.5*	*73.7*	*g N eq*
Diffuse emissions	34.79	89.47	36.72	36.46	45.34	%
Cow feed cultivation	22.79	5.08	57.05	0.67	0.18	%
Fodder production	42.42	5.45	6.04	62.77	54.35	%
Energy use	0.00	0.00	0.18	0.10	0.09	%
Agrochemical production	0.00	0.00	0.00	0.00	0.04	%
Land occupation	0.00	0.00	0.00	0.00	0.00	%
Cleaning activities	–	–	–	–	0.00	%
WD – Total	*0.55*	*0.53*	*0.27*	*0.61*	*0.68*	*m^3 water*
Diffuse emissions	0.00	0.00	0.00	0.00	0.00	%
Cow feed cultivation	2.36	2.41	8.62	1.23	0.36	%
Fodder production	49.36	4.98	13.40	59.05	56.17	%
Energy use	0.00	0.00	1.26	0.28	0.30	%
Agrochemical production	0.00	0.00	0.00	0.00	0.29	%
Water use	48.28	92.61	76.72	39.44	42.88	%
Cleaning activities	–	–	–	–	0.00	%
FD – Total	*0.65*	*0.61*	*0.59*	*0.86*	*0.69*	*kg oil eq*
Diffuse emissions	0.00	0.00	0.00	0.00	0.00	%
Cow feed cultivation	49.89	94.84	66.79	41.70	17.33	%
Fodder production	50.11	5.16	7.21	49.40	65.68	%
Energy use	0.00	0.00	25.99	8.90	13.31	%
Agrochemical production	0.00	0.00	0.00	0.00	3.68	%
Land occupation	0.00	0.00	0.00	0.00	0.00	%
Cleaning activities	–	–	–	–	0.00	%

(continued)

Table 9 (continued)

Normalized results	S1	S2	S3	S4	S5	Units
NI—Total	*0.076*	*0.063*	*0.056*	*0.087*	*0.091*	–
Diffuse emissions	6.98	15.82	8.58	5.28	7.65	%
Cow feed cultivation	15.77	28.57	30.27	12.53	4.03	%
Fodder production	52.06	6.11	9.25	59.24	60.93	%
Energy use	0.00	0.00	11.88	3.80	4.40	%
Agrochemical production	0.00	0.00	0.00	0.00	0.98	%
Land occupation	25.20	49.51	40.03	19.15	22.01	%
Cleaning activities	–	–	–	–	0.00	%

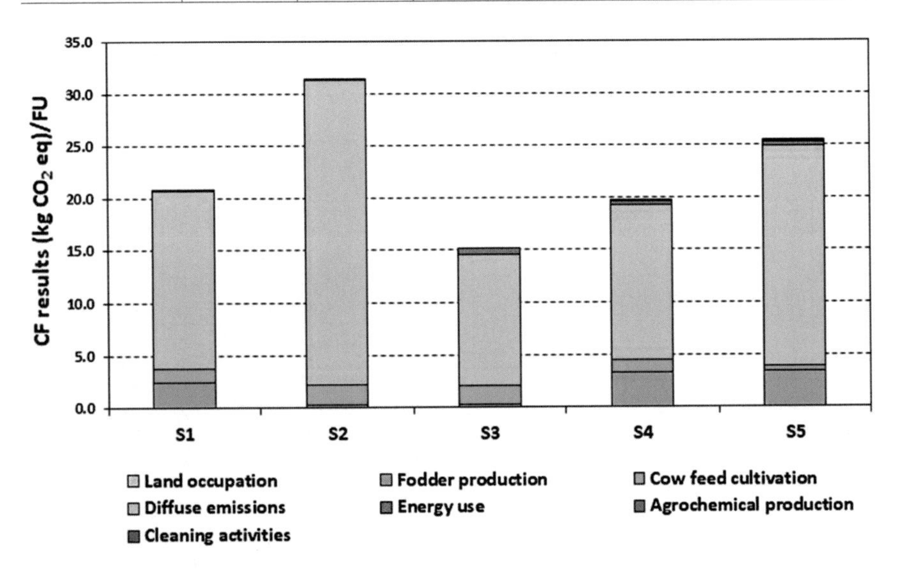

Fig. 4 CF results for the different processes involved in beef production in farming systems

3.1 Normalized Results

Focusing on the normalized results (Table 9), S3 was responsible for the lowest impacts (0.056), while S4 and S5 reported the worst results (0.087 and 0.091, respectively), followed by S1 (0.076), from an environmental perspective. This is mainly due to the key role of fodder production in these systems, since higher ratios must be supplied to calves according to the information provided by producers on calve diets. In general, the production of fodder can be an energy-intensive process, in

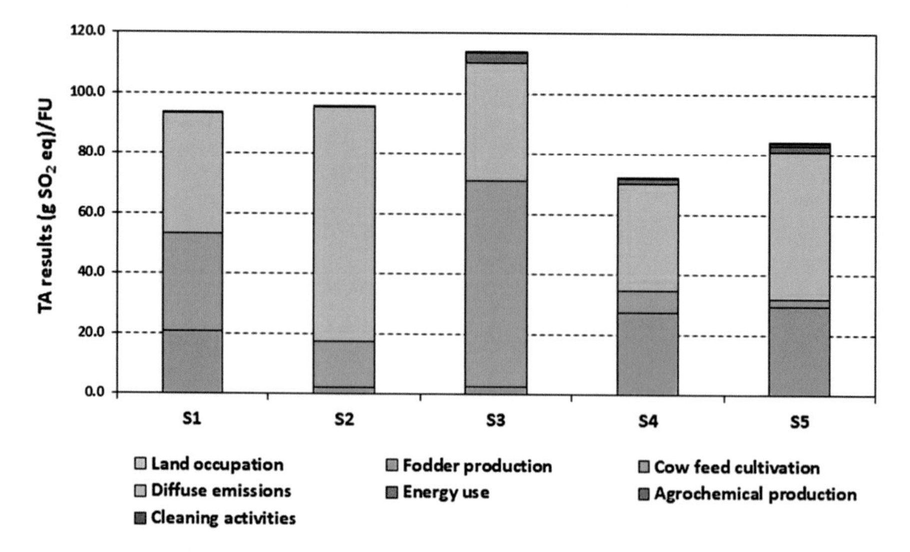

Fig. 5 TA results for the different processes involved in beef production in farming systems

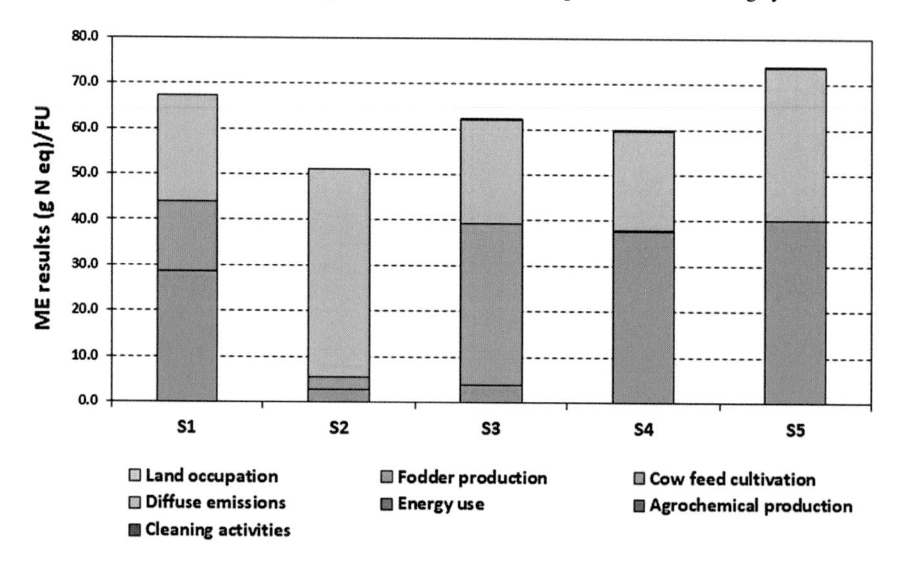

Fig. 6 ME results for the different processes involved in beef production in farming systems

addition to covering all the environmental burdens associated with the cultivation of those ingredients (mainly cereal crops) that are required for feed formulation.

In line with the above, three main contributing factors were found as major responsible for the environmental impacts (Fig. 3): cow feed cultivation, fodder production and land occupation. However, their relative contribution is variable in relation to the different producers and farming systems. Thus, feed production (grouping together

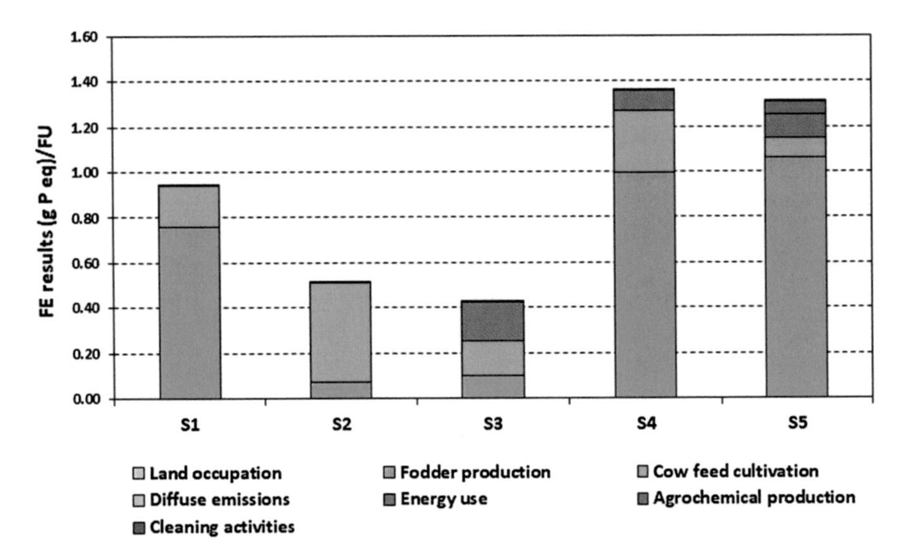

Fig. 7 FE results for the different processes involved in beef production in farming systems

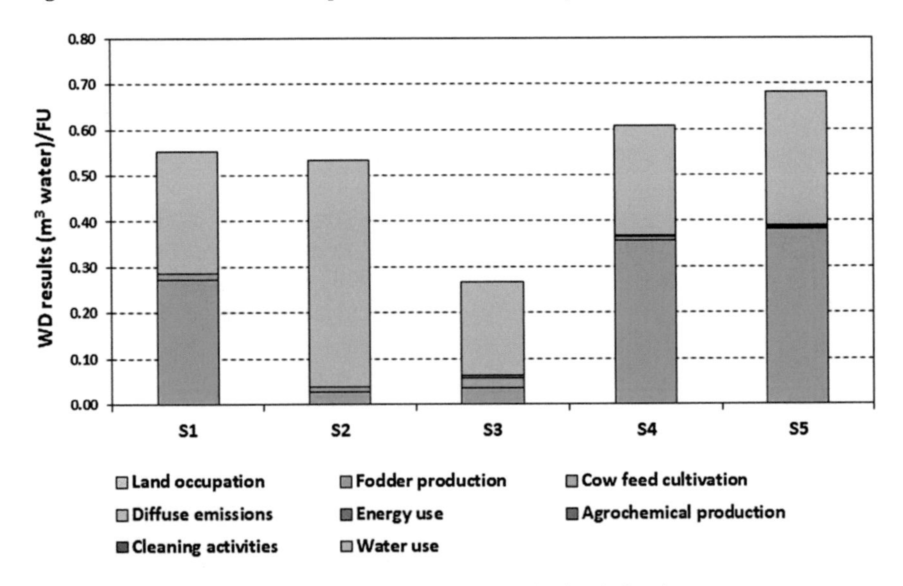

Fig. 8 WD results for the processes involved in beef production in farming systems

both feed cultivation and fodder production) generates between 35% (S2) and 72% (S4) of the impacts on this indicator. In the case of S1, S4 and S5, this impact is mainly due to the production of fodder supplied to calves, while in the case of S2 and S3 it is due to the cultivation of the dried grass and the chopped maize, respectively. These differences are directly related to the fact that the amount and proportions of feed supplied to the herd are different in each farm (see Table 2). Similarly, the land occupied by each farming system (as a whole) is also variable, contributing between 19% (S4) and 49% (S2) of the standard impacts: more agricultural land is required in S2 per kilogram of beef produced (FU), leading to greater impacts, while similar requirements were recorded in the other systems.

Diffuse emissions from farm also made a significant contribution (from 5 to 16%), mainly due to emissions of nitrogen compounds (NO_3^- and NH_3) from manure management. Their influence is especially relevant in S2, since higher ratios of manure are generated in relation to the herd size at farm compared to the other systems. On the contrary, no damages were registered from energy use in this system (S2), due to the absence of electricity requirements; the same applies to S1. Finally, agrochemical production and cleaning activities had a negligible influence on all the systems evaluated.

3.2 CC (CF) Results

As can be seen from the results obtained (Table 9; Fig. 4), S2 had the worst results for CF (31.2 kg CO_2/FU), while S3 had the lowest impacts (15.1 kg CO_2/kg FU). In all systems, diffuse emissions from farm are the main contributor to CF environmental burdens. More in detail, these emissions represent between 77% (S4) and 93% (S2) of the impacts, mainly due to the direct CH_4 emissions that occur during the enteric fermentation of cows. These emissions depend on the number of cows on the farm, so the differences in results are also directly linked to the relationship between annual beef production and the number of cows (herd size) on the farm (see Table 4). Feed production (i.e. grass and maize cultivation and fodder manufacture) also contributes to environmental impacts, but to a lesser extent (between 7% and 23% of impacts). In this case, the impacts produced are proportional to the quantity of each ingredient consumed in each farm (see Table 2), in relation to beef production, since it is the basis for comparison (FU). These impacts on feed production are due to agricultural activities carried out during cultivation, especially due to the consumption of fossil fuels such as diesel in tractors. On the contrary, a negligible effect (below 1%) was registered for the other factors, except for energy use, which has a minor impact in some farming systems (S3—S5) due to the consumption of electrical energy resulting from some farm activities (such as animal feeding).

3.3 Acidification (TA) Results

Similar to CF results, diffuse emissions and feed production played a key role in the acidification potential associated with the different beef production systems evaluated (Fig. 5). However, in this case, the relative contribution of these processes is more balanced between them. Thus, although diffuse emissions still maintain a greater effect on S2 (due to higher NH_3 emissions from higher manure generation rates), the influence of animal feed production becomes relevant in the other systems: while the cultivation of grass and maize share an important effect in S1 and S2, fodder production gains relevance in S4 and S5.

On the other hand, energy use continued to have a minor impact (around 3%) on acidification results, regardless of the system considered, as well as agrochemical production and cleaning activities (far below 1%).

3.4 Eutrophication (ME, FE) Results

Analogous to TA results, eutrophication impacts on marine environments (ME) was mainly linked to N-based emissions (up to 89% in S2), resulting from both manure management (storage and application) and cultivation practices to produce the different ingredients involved in herd diets. Consequently, the effect of the other contributing factors is barely noticeable (Fig. 6).

Conversely, while cow feed cultivation and fodder production continue to have a relevant impact on FE, diffuse emissions from manure management at farm can be considered to be virtually zero (Fig. 7). The reason is that the results of freshwater eutrophication focus on the impact of the phosphorus (P) compounds discharged to the environment. Therefore, since no P-based emissions were estimated for manure management in this study, no impacts can be associated with this factor in this impact category. This is directly related to the approach followed by the ReCiPe method to estimate the eutrophication potential divided into two different impact categories (ME and FE), unlike other previous methods (such as CML 2001 method—Guinée et al. 2001) that combine all eutrophication impacts in a single category, expressed as equivalent emissions of phosphate (PO_4^{3-}) or NO_3^-, depending on the method. In this way, all impacts derived from direct and indirect nitrogen emissions are directly attributed to the ME results, giving higher scores to diffuse emissions (Fig. 6).

3.5 WD Results

As expected, direct water consumption was identified as the main contributor to the impacts on water resource depletion, primarily related to the continued supply of water to animals in the different farming systems (Fig. 8). However, indirect consumption also had a significant influence on the results, especially in terms of fodder production, due to the higher requirements of water for the cultivation of the cereals (maize, barley, rye and soybean).

Regarding the production of the other ingredients of the herd diets, minor water application is required for the on-farm cultivation of maize, while irrigation is not necessary for dried grass. These low levels of water consumption could be closely related to the weather conditions in the target area under study. Galicia has an Atlantic climate with heavy rainfall for most of the year and a small dry season. In this sense, the official reports on this region point especially to the humid springs of recent years, following the trend of previous months, with a number of rainy days that exceed the expected climate values (MeteoGalicia 2016).

4 WF Results

Table 10 includes the WF results (per FU) obtained for the different systems evaluated, as well as the relative contribution of both SS1 and SS2 to the overall results. Consequently, animal feed (SS1) represented the largest impact (over 95%), directly related to the use of water from agricultural activities associated with the cultivation of grass and cereals; therefore, it stands as the most critical process (commonly named *hotspot*) of the global system. On the other hand, the subsystem involving the management of the farm (SS2) had a minor relevance (up to 5%); in this case, the WF impact was mainly related to the direct demand for water for the herd, as well as to the indirect consumption of water from the use of energy on the farm and the transport of inputs from their origin to the farm facilities.

Table 10 also presents disaggregated results related to green, blue and grey WF. Focusing on SS1, green WF was found to be primarily responsible for the critical role of feed production, with about 50% of the impacts, due to evapotranspiration of rainwater from agricultural soils, especially in the case of rye cultivation (up to 44%).

Table 10 Global and disaggregated (green, blue, grey) WF results per FU (1 kg of carcass beef at farm gate) for the different systems and subsystems

System/Subsystem	WF results				Units
	Green WF	Blue WF	Grey WF	Total WF	
S1	*3.15*	*2.55*	*0.97*	*6.67*	*m³ water*
S1–SS1	100	90.0	100	96.2	%
S1–SS2	0.00	10.0	0.00	3.80	%
S2	*0.33*	*0.71*	*0.10*	*1.14*	*m³ water*
S2–SS1	100	33.1	100	58.7	%
S2–SS2	0.00	66.9	0.00	41.3	%
S3	*2.25*	*1.20*	*0.81*	*4.26*	*m³ water*
S3–SS1	100	82.2	100	95.0	%
S3–SS2	0.00	17.8	0.00	5.00	%
S4	*3.04*	*2.85*	*0.86*	*6.75*	*m³ water*
S4–SS1	100	91.6	100	96.4	%
S4–SS2	0.00	8.4	0.00	3.60	%
S5	*3.24*	*3.07*	*0.92*	*7.23*	*m³ water*
S5–SS1	100	90.5	100	96.0	%
S5–SS2	0.00	9.50	0.00	4.00	%

However, blue WF due to water depletion of surface or groundwater courses had a lower contribution (from 24.5 to 40.0%), followed closely by grey WF (below 20%). In contrast, the impact of WF on SS2 was entirely derived from the blue component, while grey WF had a lower contribution (below 1%); green WF had no effect on the results.

The only exception occurs in S2, with much lower global WF impacts, as well as more equitable contributions between SS1 and SS2. The rationale behind these disparities is directly associated with both green and blue components. Green WF is mainly related to the cultivation of the ingredients to meet the demand for animal feed at farm. In S2, the consumption of dried grass predominated over other dietary inputs, so that the environmental burdens coming from its production were responsible for the greatest impacts.

However, in the absence of reliable primary data, approximate impact factors were applied to grass in this study, which may have led to an underestimation of the related impacts. On the other hand, the blue WF depends on the water extracted from surface or groundwater resources, so that higher water demand led to higher blue WF impacts. In this regard, the ratio of water consumption in S2 (494 L water/kg of carcass beef) was higher than in the other systems (less than 300 L water/kg of carcass beef), so that water demand at farm gained importance in the overall results. The volume of manure generated by the cattle herd in S2 was also higher (25 kg cow manure/kg of carcass beef), which means that a large part of the water consumed by the animals can be returned to the environment through the application of manure to soils; however, it was not enough to offset the related impacts in SS2, so that they became comparable to SS1 (Table 10).

5 Discussion

5.1 WF Versus WD

As aforementioned, both WD and WF results were estimated in the present study with the aim of analyzing the degree of agreement among them. According to the results, much lower values were reported for the WD results (Table 9) compared to their analogues in terms of WF impact (Table 10). The rationale behind this lies on the different methodological approaches followed in both cases. The WD concept only covers the environmental impacts due to the water used as an input to the system; on the contrary, WF also includes the impact from the water evaporated or uptaken by the growing plants, along with the water that will be added to the environment to meet the water quality standards defined by the competent authorities. In this way, the WD results could be understood as a fraction of the WF impact, which provides a more comprehensive approach involving water usage and related impacts (Hoekstra et al. 2011).

Table 11 Comparative assessment between WD and WF results: relevance of green WF impact

System	WD results (m^3 water)	WF results			WD relative to WF (%)
		Total WF (m^3 water)	Green WF (m^3 water)	Green WF (%)	
S1	0.55	6.67	3.15	47.2	8.25
S2	0.53	1.14	0.33	28.9	46.5
S3	0.27	4.26	2.25	52.8	6.34
S4	0.61	6.75	3.04	45.0	9.04
S5	0.68	7.23	3.24	2.23	9.41

This effect is particularly relevant in those systems where agricultural activities for animal feed production play a critical role, in which green WF exerts the greatest influence. Accordingly, Table 11 shows a comparative assessment between WD and WF results to prove that differences increase as the green component enhance its contribution to the overall WF results. Only S2 breaks the tendency in line with the predominance of the blue WF in this system compared to the others.

5.2 *Benchmarking with Similar WF Studies in the Literature*

As mentioned at the beginning of the chapter, there are far fewer studies available in the current literature on beef production and its WF indicator, especially within European borders, compared to more common environmental research. However, similarly to the previous comparative analysis with WD results, the estimated WF values for the Galician Natural Parks were also compared with those reported by authors from other countries. The average WF ratio in the present study was estimated for this purpose; however, it should be noted that S2 results were not taken into consideration due to their variance as compared to the other four systems.

A summary of the studies selected for comparison is reported in Table 12. According to the results, close WF values were registered in the most current beef production systems in other European countries (i.e. The Netherlands), about 6 m^3 of water per kg of beef product. WF ratios increase, however, in previous studies on European beef products, becoming aware of less efficient water opportunities (Hoekstra and Chapagain 2007). Similarly, beyond European borders, much greater impacts on water resources have been also registered, with values ranging from 13.7 to 33.3 kg m^3 of water/kg of beef at farm gate. Accordingly, Molden et al. (2007) ensure that levels of water productivity can be highly dependent on the type of product as well as the area where it is produced. In effect, those regions with extreme poverty and dependence of agriculture (mainly in Africa and South Asia) show the lowest yields in terms of kg of product per m^3 of water consumed; special attention could be paid to these areas, therefore, to optimize water usage and contribute to reduce poverty worldwide

Table 12 WF results per FU (1 kg of carcass beef at farm gate) of the beef production systems selected for comparison

Study	Country	WF results (m^3 water)
This study	Spain	6.23
European countries		
Hoekstra and Chapagain (2007)	Italy	21.2
	The Netherlands	11.7
Mekonnen and Hoekstra (2012)	The Netherlands	6.51
Other foreign countries		
Renault and Wallender (2000)	USA	13.5
Hoekstra and Chapagain (2007)	World average	15.5
	USA	13.2
	India	16.5
	China	12.6
	Russia	21.0
	Indonesia	14.8
	Australia	17.1
	Brazil	17.0
	Japan	11.0
	Mexico	37.8
Molden et al. (2007)	World average	10.0–33.3
Mekonnen and Hoekstra (2012)	USA	14.2
	India	16.6
	China	13.7
Peters et al. (2010)	Australia	17.1
Ridoutt et al. (2014)	Australia	0.07

(Molden et al. 2007). Otherwise, Ridoutt et al. (2012, 2014) reported extremely low results in relation to beef production in Australia; this contrasts not only with the WF results calculated in other countries, but also with the values published by other authors in the same region (Hoekstra and Chapagain 2007; Peters et al. 2010). However, the disparities in this case can be directly attributed to the analysis approach, since Ridoutt et al. (2012, 2014) use normalized results (instead of characterization ones) to express impacts on water use.

In any case, these outcomes would be in line with the great variability in the WF results highlighted by various review studies in the field (Aivazidou et al. 2016; Gerbens-Leenes et al. 2013). According to them, the efficiency of food conversion, together with the composition and origin of animal feed, can greatly affect the WF impacts associated with beef production, similar to those of other meat products on the market. Likewise, the methodologies used in the estimations may also have a significant influence on the results (Aivazidou et al. 2016), as has recently been demonstrated in the present and previous sections (WF vs WD analysis).

5.3 Benchmarking with Similar CF Studies in Literature

Several studies are available in the literature on the assessment of beef production from an environmental perspective (de Vries et al. 2015); however, only some of them are developed within European borders (Mogensen et al. 2015; Nguyen et al. 2010; Opio et al. 2013). Similarly, most environmental studies in Spain focus on dairy systems rather than beef production, making the comparative analysis more limited with the results obtained in this study. Moreover, most of the LCA studies on beef productive chains evaluate only the CF impacts, while other environmental indicators such as acidification and eutrophication potentials are left out of the system boundaries; in this line, no studies on the impact of the depletion of water resources were found. With this in mind, Table 12 summarizes the main characteristics (country, farming regime, animal diet and characterization results) of the studies selected for comparison in terms of CF results following the LCA perspective. It should be noted that the mean values were calculated on the basis of the comparison; again CF results in S2 were not considered due to their variance in comparison with the other systems.

According to the results, beef production in European countries was found to share CF values of the same order of magnitude as those reaches in this study within the framework of the Galician parks (20.1 vs 23.1–31.0 kg CO_2 eq/kg of carcass beef), although just superior. This could be not only related to the farming regime but also with the type of ingredients supplied to the cattle herd. According to literature, those systems prioritizing forage-based diets accounted for the highest CF impacts, as a rule, compared to fodder-based alternatives; however, it should also be highlighted that minor differences were reported. In addition, diffuse emissions were also identified as one of the factors that contribute most to the overall impact of CF, along with animal feed.

Similar CF results were registered in other foreign countries outside Europe, which again exceed the values estimated for the Galician systems, regardless of the type of farming regime and priority animal diet. Similarly, focusing on the same system, diets composed mainly of fodder remain responsible for the least environmentally friendly outcomes. The same applies to extensive systems compared to intensive regimes (Table 13).

On the basis of the above, it can be concluded that, despite extensive practices, the use of balanced diets and good agricultural practices makes Galician systems a reliable example on the road to environmental sustainability in the livestock sector worldwide.

Table 13 Main characteristics and CF results per FU (1 kg of carcass beef at farm gate) of the beef production systems selected for comparison

Study	Country	Regime	Animal diet	CF results (kg CO_2 eq)
This study	*Spain*	*Extensive*	*Forage/Fodder*	*20.1*
European countries				
Mogensen et al. (2015)	Denmark	Intensive	Forage	23.1
	Sweden	Intensive	Forage	25.4
Nguyen et al. (2010)	European Union	Intensive	Forage	27.3
Opio et al. (2013)	Western Europe	N.D.	N.D.	31.0
	Eastern Europe	N.D.	N.D.	29.1
Other foreign countries				
Alig et al. (2012)	Switzerland	Intensive	Forage	27.5[a]
Beauchemin et al. (2010)	Canada	Intensive	Fodder	21.7
		Intensive	Forage	23.1
Cederberg et al. (2009)	Brazil	Intensive	N.D.	28.0
Lupo et al. (2013)	USA	Intensive	Fodder	23.0
		Intensive	Forage	31.5
		Extensive	Forage	26.6[a]
Opio et al. (2013)	North America	N.D.	N.D.	35.2
Pelletier et al. (2010)	USA	Intensive	Fodder	26.6[a]

[a]Re-calculated for 1 kg of carcass beef assuming 1 kg of live weight beef is equivalent to 1.8 kg of carcass beef
N.D.: no data available

5.4 *Certification Procedure*

The eco-label is awarded when the products evaluated are obtained through both traditional and good practices. These practices should include, with respect to conventional production, energy saving, water saving, reduction of the use of chemicals such as detergents and disinfectants and reduction of waste generated. For this reason, the maximum permissible environmental impact thresholds for certification are identified, below which it is ensured that production is carried out in an environmentally friendly manner.

The European Ecological Label (EEA; Eco-label), which is considered one of the instruments included in the EU Sustainable Consumption and Production and Sustainable Industrial Policy Action Plan, is responsible for defining the procedure for establishing and accrediting the Ecolabel for various products. Thus, the range of products analysed in the EEA ranges from cleaning products, household appliances, textiles, lubricants, paints and varnishes to accommodation and camping facilities.

In this context, the procedure followed in this study to develop the Ecolabel for the products assessed (beef) shares the philosophy and procedures with the EEA. However, it should be noted that it has been specifically designed for the evaluation of products in the food sector. The fact of analyzing all the consumptions and impacts associated with the life cycle allows a quantitative evaluation of them, in a way that complements the recommendations established as good practices, but allows comparison with other products of a similar nature, provided that a life cycle study of these products has been carried out to obtain their environmental indicators.

According to the main findings of this study, the following common criteria were established as a requirement for obtaining the eco-label in beef production systems within the Community of Natural Parks in Galicia:

1. Comply with the guide to good beef production practices.
2. Have a NI of less than 0.08/kg beef (carcass weight).
3. Have a CF impact of less than 25 kg CO_2 eq/kg beef (carcass weight).
4. Have a WF impact of less than 5 m^3 water/kg beef (carcass weight).

6 Conclusions

In recent times, society has faced problems related to food security and environmental damage. As a primary source of protein and micronutrients, meat varieties and related products are considered essential elements of the human diet. More specifically, beef stand as one of the meat varieties with larger consumption ratios worldwide. However, beef production is also responsible for significant consequences—both directly and indirectly—on the environment and its preservation. This is why consumers are increasingly demanding an evolution towards the generation of more sustainable productive sector involving beef products with reduced environmental impacts.

In this context, the present study aimed to carry out the environmental assessment of different systems involving beef production within Natural Parks in Galicia, following the principles established by the LCA methodology in combination with the WFN guidelines. Based on the results obtained, it can be concluded that the systems evaluated are environmentally friendly, based on the ratios associated with the indicators evaluated and their favourable comparison with similar studies in other foreign regions published to date on the beef sector.

Finally, the main outcomes of the study served as a basis for establishing the methodological framework for the implementation of an ecological certification as a distinctive sign of quality in the market for beef products from Galician Natural Parks. To this aim, the premises of a sustainable and environmentally responsible production process were defined, paying special attention to the impacts related to Carbon and Water footprints.

Acknowledgements This study was carried out within the framework of the research work "Investigación en los parques naturales sobres captación de CO_2 y N_2O por diferentes cultivos para contribuir a mitigar el cambio climático". Project funded by Fundación La Caixa and Xunta de Galicia. The authors (I. Noya, L. Lijó, G. Feijoo and M.T. Moreira) belong to CRETUS (AGRUP2015/2) and the Galician Competitive Research Group GRC ED431C 2017/29, programme co-funded by Xunta de Galicia and FEDER.

References

Alexandratos, N., & Bruinsma, J. (2012). World Agriculture Towards 2030/2050: The 2012 Revision. FAO, Rome, ESA Working Paper No. 12-03.

Aivazidou, E., Tsolakis, N., Iakovou, E., & Vlachos, D. (2016). The emerging role of water footprint in supply chain management: a critical literature synthesis and a hierarchical decision-making framework Review. *Journal of Cleaner Production, 137,* 1018–1037. https://doi.org/10.1016/j.jclepro.2016.07.210.

Alig, M., Grandl, F., Mieleitner, J., Nemecek, T., & Gaillard, G. (2012). *Ökobilanzvon Rind-, Schweine- und Geflügelfleisch, Schlussbericht September 2012.* Zürich, Switzerland: Agroscope Reckenholz-Tänikon ART.

Althaus, H. J., Chudacoff, M., Hischier, R., Jungbluth, N., Osses, M., & Primas, A. (2007). Life cycle inventories of chemicals. Ecoinvent Report No. 8, v2.0 EMPA. Swiss Centre for Life Cycle Inventories, Dübendorf, Switzerland.

Beauchemin, K. A., Janzen, H. H., Little, S. M., McAllister, T. A., & McGinn, S. M. (2010). Life cycle assessment of greenhouse gas emissions from beef production in western Canada: A case study. *Agricultural Systems, 103*(6), 371–379. https://doi.org/10.1016/j.agsy.2010.03.008.

Beef Market Central. (2016). World beef consumption per capita (ranking of countries). http://beef2live.com/story-world-beef-consumption-per-capita-ranking-countries-0-111634.

Cederberg, C., Meyer, D., & Flysjö, A. (2009). Life cycle inventory of greenhouse gas emissions and use of land and energy in Brazilian beef production. SIK Report no. 792. The Swedish Institute for Food and Biotechnology.

Cederberg, C., Persson, U. M., Neovius, K., Molander, S., & Clift, R. (2011). Including carbon emissions from deforestation in the carbon footprint of Brazilian beef. *Environmental Science and Technology, 45*(5), 1773–1779. https://doi.org/10.1021/es103240z.

Chapagain, A. K., & Hoekstra, A. Y. (2004). Water footprints of nations. Value of Water Research Report Series No. 16. http://waterfootprint.org/media/downloads/Report16Vol1.pdf.

Ciolos, D. (2012). European Commissioner for Agriculture and Rural Development. European Union. http://europa.eu/rapid/press-release_SPEECH-12-480_en.htm.

Daneshi, A., Esmaili-sari, A., Daneshi, M., & Baumann, H. (2014). Greenhouse gas emissions of packaged fluid milk production in Tehran. *Journal of Cleaner Production, 80,* 150–158. https://doi.org/10.1016/j.jclepro.2014.05.057.

De Léis, C. M., Cherubini, E., Ruviaro, C. F., da Silva, V. P., Lampert, V. N., Spies, A., et al. (2015). Carbon footprint of milk production in Brazil: a comparative case study. *International Journal of Life Cycle Assessment, 20,* 6–60. https://doi.org/10.1007/s11367-014-0813-3.

De Vries, M., & de Boer, I. J. M. (2010). Comparing environmental impacts for livestock products: a review of life cycle assessments review. *Livestock Science, 128,* 1–11. https://doi.org/10.1016/j.livsci.2009.11.007.

De Vries, M., van Middelaar, C. E., & de Boer, I. J. M. (2015). Comparing environmental impacts of beef production systems: a review of life cycle assessments. *Livestock Science, 178,* 279–288. https://doi.org/10.1016/j.livsci.2015.06.020.

Directive 91/271/EEC. (1991). Council directive 91/271/EEC of 21 May 1991 concerning urban wastewater treatment. Office Journal European Communities 40–52 (No L 135, 30/05/1991).

Dones, R., Bauer, C., Bolliger, R., Burger, B., Faist Emmenegger, M., & Frischknecht, R. et al. (2007). Life Cycle Inventories of Energy Systems: Results for Current Systems in Switzerland and Other UCTE Countries. Ecoinvent Report No. 5. Paul Scherrer Institut Villigen, Swiss Centre for Life Cycle Inventories, Dübendorf, Switzerland.

Fantin, V., Buttol, P., Pergreffi, R., & Masoni, P. (2012). Life cycle assessment of Italian high quality milk production. A comparison with an EPD study. *Journal of Cleaner Production, 28,* 150–159. https://doi.org/10.1016/j.jclepro.2011.10.017.

FAO. (2014). *Global Warming: the Impact of Meat Sector in the European Union*. Brussels: Directorate-General for Agriculture.

FAO. (2006). *Livestock's long shadow environmental issues and options* (p. 390). Rome: Food and Agriculture Organization of the United Nations. http://www.fao.org/docrep/010/a0701e/a0701e00.HTM.

Franke, N. A., Boyacioglu, H., & Hoekstra, A. Y. (2013). Grey water footprint accounting: Tier 1 supporting guidelines. Value of Water Research Report Series No. 65, UNESCO-IHE, Delft, the Netherlands.

Galloway, J. N., Burke, M., Bradford, G. E., Naylor, R., Falcon, W., & Chapagain, A. K. et al. (2007). International Trade in Meat: The Tip of the Pork Chop. *Ambio, 36*(8), 622–629. https://doi.org/10.1579/0044-7447(2007)36%5b622:ITIMTT%5d2.0.CO;2.

Gerbens-Leenes, P. W., Mekonnen, M. M., & Hoekstra, A. Y. (2013). The water footprint of poultry, pork and beef: a comparative study in different countries and production systems. *Water Resources and Industry, 1–2,* 25–36. https://doi.org/10.1016/j.wri.2013.03.001.

Gerber, P. J., Steinfeld, H., Henderson, B., Mottet, A., Opio, C., & Dijkman, J. (2013). *Tackling Climate Change through Livestock: A Global Assessment of Emissions and Mitigation Opportunities*. Rome: Food and Agriculture Organization of the United Nations (FAO).

Gerber, P. J., Mottet, A., Opio, C. I., Falcucci, A., & Teillard, F. (2015). Environmental impacts of beef production: Review of challenges and perspectives for durability. *Meat Science, 109,* 2–12. https://doi.org/10.1016/j.meatsci.2015.05.013.

Goedkoop, M., Oele, M., Leijting, J., Ponsioen, T., & Meijer, E. (2013a). *Introduction to LCA with SimaPro 8*. The Netherlands: PRé Consultants.

Goedkoop, M., Heijungs, R., Huijbregts, M., De Schryver, A., Struijs, J., & Van Zelm, R. (2013b). ReCiPe 2008. In *A Life Cycle Impact Assessment Method Which Comprises Harmonised Category Indicators at the Midpoint and the Endpoint Level; Report I: Characterisation* (1st ed.). http://www.lcia-recipe.net.

González-García, S., Castanheira, E. G., Dias, A. C., & Arroja, L. (2013). Using life cycle assessment methodology to assess UHT milk production in Portugal. *Science of the Total Environment, 442,* 225–234. https://doi.org/10.1016/j.scitotenv.2012.10.035.

Guinée, J. B., Gorrée, M., Heijungs, R., Huppes, G., Kleijn, R., de Koning, A., et al. (2001). *Life Cycle Assessment—An Operational Guide to the ISO Standards* (p. 2001). The Netherlands: Centre of Environmental Science, Leiden.

Harding, G., Courtney, C., & Russo, V. (2017). When geography matters. A location-adjusted blue water footprint of commercial beef in South Africa. *Journal of Cleaner Production, 151,* 494–508. https://doi.org/10.1016/j.jclepro.2017.03.076.

Hoekstra, A. Y., & Chapagain, A. K. (2007). Water footprints of nations: water use by people as a function of their consumption pattern. *Water Resources Management, 21*(1), 35–48. https://doi.org/10.1007/s11269-006-9039-x.

Hoekstra, A. Y., Chapagain, A. K., Aldaya, M. M., & Mekonnen, M. M. (2011). *The Water Footprint Assessment Manual*. Earthscan, London: Setting the Global Standard.

INTERAL. (2008). *Estudio de posicionamiento estratégico para el sector de alimentación animal en el escenario actual* (p. 93). Organización Interprofesional Española de la Alimentación Animal: INTERAL.

ISO 14040. (2006). *Environmental Management—Life Cycle Assessment—Principles and Framework*. Geneva, Switzerland: International Organization of Standardization.

ISO 14044. (2006). *Environmental Management—Life Cycle Assessment—Requirements and Guidelines*. Geneva, Switzerland: International Organization of Standardization.
ISO 14067. (2013). *Carbon Footprint of Products—Requirements and Guidelines for Quantification and Communication*. Geneva, Switzerland: International Organization of Standardization.
IPCC 2006. (2006). IPCC guidelines for national greenhouse gas inventories. Intergovernmental panel on climate change. In Eggleston, H. S., Buendia, L., Miwa, K., Ngara, T., & Tanabe, K. (Eds.), *National Greenhouse Gas Inventories Program*. IGES, Japan, ISBN: 4-88788-032-4.
IPCC. (2013). *Climate Change 2013: The physical science basis. Contribution of Working Group I to the Fifth Assessment Report of the Intergovernmental Panel on Climate Change*. In Stocker, T. F., Qin, D., Plattner, G.-K., Tignor, M., Allen, S. K., & Boschung, J. (Eds.). (pp. 1535) Cambridge University Press, Cambridge, United Kingdom and New York, NY, USA.
Lupo, C. D., Clay, D. E., Benning, J. L., & Stone, J. J. (2013). Life-cycle assessment of the beef cattle production system for the Northern Great Plains, USA. *Journal of Environmental Quality Abstract—Environmental Models, Modules, and Datasets, 42*(5), 1386–1394. https://doi.org/10.2134/jeq2013.03.0101.
MAPAMA. (2017). Ministerio de Agricultura y Pesca, Alimentación y Medio Ambiente. Sección Ganadería. http://www.mapama.gob.es/es/ganaderia/temas/ganaderia-y-medio-ambiente/.
McAuliffe, G. A., Chapman, D. V., & Sage, C. L. (2016). A thematic review of life cycle assessment (LCA) applied to pig production. *Environmental Impact Assessment Review, 56,* 12–22. https://doi.org/10.1016/j.eiar.2015.08.008.
McClelland, S. C., Arndt, C., Gordon, D. R., & Thoma, G. (2018). Type and number of environmental impact categories used in livestock life cycle assessment: A systematic review. *Livestock Science, 209,* 39–45. https://doi.org/10.1016/j.livsci.2018.01.008.
MeteoGalicia. (2016). Informe climatolóxico primavera 2016. Consellería de Medio Ambiente e Ordenación do Territorio. Secretaría Xeral de Calidade e Avaliación Ambiental. Xunta de Galicia.
Mekonnen, M. M., & Hoekstra, A. Y. (2010). The green, blue and grey water footprint of crops and derived crop products. Value of Water Research Report Series No. 47. The Netherlands: UNESCOIHE, Delft.
Mekonnen, M. M., & Hoekstra, A. Y. (2012). A global assessment of the water footprint of farm animal products. *Ecosystems, 15,* 401–415. https://doi.org/10.1007/s10021-011-9517-8.
Mogensen, L., Kristensen, T., Nielsen, N. I., Spleth, P., Henriksson, M., Swensson, C., et al. (2015). Greenhouse gas emissions from beef production systems in Denmark and Sweden. *Livestock Science, 174,* 126–143. https://doi.org/10.1016/j.livsci.2015.01.021.
Molden, D., Oweis, T. Y., Steduto, P., Kijne, J. W., Hanjra, M. A., & Bindraban, P. S. et al. (2007). Pathways for increasing agricultural water productivity. In Molden, D. (Ed.), *Water for food, water for life: A comprehensive assessment of water management in agriculture* (pp. 279–310). London, Colombo: Earthscan, International Water Management Institute.
Nemecek, T., & Käggi, T. (2007). Life Cycle Inventories of Agricultural Production Systems. Final Report Ecoinvent v2.0 No. 15a. Agroscope FAL Reckenholz and FAT Taenikon. Swiss Centre for Life Cycle Inventories, Zurich and Dübendorf, Switzerland.
Nguyen, T. L. T., Hermansen, J. E., & Mogensen, L. (2010). Environmental consequences of different beef production systems in the EU. *Journal of Cleaner Production, 18,* 756–766. https://doi.org/10.1016/j.jclepro.2009.12.023.
Notarnicola, B., Hyashi, K., Curran, M. A., & Huisingh, D. (2012). Progress in working towards a more sustainable agri-food industry. *Journal of Cleaner Production, 28,* 1–8. https://doi.org/10.1016/j.jclepro.2012.02.007.
OECD/FAO. (2017). *OECD-FAO Agricultural Outlook 2017–2026*. Paris: OECD Publishing. http://dx.doi.org/10.1787/agr_outlook-2017-en.
Opio, C., Gerber, P., Mottet, A., Falcucci, A., Tempio, G., & MacLeod, M. et al. (2013). *Greenhouse Gas Emissions from Ruminant Supply Chains—A Global Life Cycle Assessment*. Rome: Food and Agriculture Organization of the United Nations (FAO).

Pelletier, N., Pirog, R., & Rasmussen, R. (2010). Comparative life cycle environmental impacts of three beef production strategies in the Upper Midwestern United States. *Agricultural Systems, 103,* 380–389. https://doi.org/10.1016/j.agsy.2010.03.009.

Peters, G. M., Wiedemann, S. G., Rowley, H. V., & Tucker, R. W. (2010). Accounting for water use in Australian red meat production. *International Journal of Life Cycle Assessment, 15,* 311–320. https://doi.org/10.1007/s11367-010-0161-x.

Pimentel, D., Berger, B., Filiberto, D., Newton, M., Wolfe, B., Karabinakis, E., et al. (2004). Water resources: Agricultural and environmental issues. *BioScience, 54*(10), 909–918. https://doi.org/10.1641/0006-3568(2004)054%5b0909:WRAAEI%5d2.0.CO;2.

PRé Consultants. (2016). *SimaPro Database Manual—Methods library*. The Netherlands.

Renault, D., & Wallender, W. W. (2000). Nutritional water productivity and diets. *Agricultural Water Management, 45*(3), 275–296. https://doi.org/10.1016/S0378-3774(99)00107-9.

Ridoutt, B. G., Sanguansri, P., Freer, M., & Harper, G. S. (2012). Water footprint of livestock: comparison of six geographically defined beef production systems. *International Journal of Life Cycle Assessment, 17,* 165–175. https://doi.org/10.1007/s11367-011-0346-y.

Ridoutt, B. G., Page, G., Opie, K., Huang, J., & Belloti, W. (2014). Carbon, water and land use footprints of beef cattle production systems in southern Australia. *Journal of Cleaner Production, 73,* 24–30. https://doi.org/10.1016/j.jclepro.2013.08.012.

Rodríguez Casado, R., Novo, P.,& Garrido, A. (2009). La huella hídrica de la ganadería española. Papeles de Agua Virtual (PAV), no 4, first ed. Madrid, Spain: Fundación Marcelino Botín.

Rodríguez Pascual, M. (2008). La ganadería extensiva en España. http://www.ruralnaturaleza.com/texto-la-ganaderia-extensiva-en-espana.

Roy, P., Nei, D., Orikasa, T., Xu, Q., Okadome, H., Nakamura, N., et al. (2009). A review of life cycle assessment (LCA) on some food products. *Journal of Food Engineering, 90,* 1–10. https://doi.org/10.1016/j.jfoodeng.2008.06.016.

Suh, S., Weidema, B., Hoejrup Schmidt, J., & Hejungs, R. (2010). Generalized make and use framework for allocation in life cycle assessment. *Journal of Industrial Ecology, 14*(2), 335–353. https://doi.org/10.1111/j.1530-9290.2010.00235.x.

Steinfeld, H., Gerber, P., Wassenaar, T., Castel, V., Rosales, M., & de Haan, C. (2006). *Livestock's long shadow: Environmental issues and options*. Rome, Italy: FAO—Food and Agriculture Organization of the United Nations.

Thornton, P. K. (2010). Livestock production: recent trends, future prospects. *Philosophical Transactions of the Royal Society of London Series B, Biological Sciences, 365*(1554), 2853–2867. https://doi.org/10.1098/rstb.2010.0134.

Valverde, C. (2017). Record year for Spanish swine and cattle production. Global Agricultural Information Network (GAIN) Report (SP1709). USDA Foreign Agricultural Service.

Van Breugel, P., Herrero, M., van de Steeg, J., & Peden, D. (2010). Livestock water use and productivity in the Nile Basin. *Ecosystems, 13*(2), 205–221. https://doi.org/10.1007/s10021-009-9311-z.

Vasilaki, V., Katsou, V., Ponsá, S., & Colón, J. (2016). Water and carbon footprint of selected dairy products: a case study in Catalonia. *Journal of Cleaner Production, 139,* 504–516. https://doi.org/10.1016/j.jclepro.2016.08.032.

Wernet, G., Bauer, C., Steubing, B., Reinhard, J., Moreno-Ruiz, E., & Weidema, B. (2016). The ecoinvent database versión 3 (part I): overview and methodology. *International Journal of Life Cycle Assessment, 21,* 1218–1230. https://doi.org/10.1007/s11367-016-1087-8.

Xunta de Galicia. (2018). Turismo de Galicia—"Galicia el buen camino". http://www.turismo.gal/que-visitar/espazos-naturais/parques-naturais/serra-da-encina-da-lastra.

Water Footprint and Consumer Products

Ignacio Cazcarro and Iñaki Arto

Abstract Water footprints of specific crops, animal, food products and forest products may typically be better captured by the study of the chains of these products with techniques under frameworks such as life cycle assessment (LCA), given the great heterogeneity in water intensities among most of these categories. Apart from these type of studies, with process analysis/specific supply chains view, also studies on Water Footprinting have been developed making use of more top-down techniques and analyses, such as extended environmental input-output (IO) models. In this regard, we examine these results making use of global multiregional IO (MRIO) databases such as World Input-Output Database (WIOD) and EXIOBASE. With them we can quantify the water footprints (WFs) of production and WFs of consumption of all (somehow aggregated) consumer products for different years in the period of 1995–2009. Results can be disaggregated by sectors and consumption categories, and compared with those results being obtained from the process analysis types of techniques. This lead us to characterize the appropriateness of each methods depending on the types of consumer products, considering also the type of supply chains up to the consumers, the boundary conditions established, etc. In particular MRIOs may suffer from aggregation errors, but also in an increasingly interconnected and globalized world they may have a role for WFs, especially to get industrial and even services ones, particularly for the computations of blue and grey water. Also avenues for integration of methods and open and future lines of research are discussed.

Keywords Water footprint · Multi-regional input-output models
Consumer products · WIOD

I. Cazcarro (✉)
Department of Economic Analysis, ARAID-Aragonese Agency
for Research and Development, Agrifood Institute of Aragon,
University of Zaragoza, Zaragoza, Spain
e-mail: ignacio.cazcarro@bc3research.org; vicenprice@gmail.com

I. Cazcarro · I. Arto
BC3-Basque Centre for Climate Change – Klima Aldaketa Ikergai,
Bilbao, Spain

S. S. Muthu (ed.), *Environmental Water Footprints*, Environmental Footprints
and Eco-design of Products and Processes,
https://doi.org/10.1007/978-981-13-2508-3_3

1 Introduction

In this chapter, we elaborate on the Water footprint[1] of consumer products, which can be defined as merchandises or commodities which are produced and subsequently consumed by individuals or households for private consumption.

More specifically, the water footprint of a product was defined (Hoekstra et al. 2009, 2011) as the volume of freshwater used to produce the commodity, measured over the full supply chain. Being a multidimensional indicator, it reveals water consumption volumes by source as well as polluted volumes by type. Components of total water footprint are specified temporally and geographically. Blue water footprint refers to the blue water resources (surface and groundwater) loss from the available ground water and surface water in a catchment area, occurring along the supply chain of a product. Losses occur when water either evaporates, or is incorporated into a product or returns to another catchment area or the sea. Green water footprint denotes consumption of green water resources, i.e., rainwater which does not become run-off. Grey water footprint is defined as the volume of freshwater required to assimilate loads of pollutants given a certain natural concentration and considering ambient water quality standards which exist.

Different methods have been proposed to consider water pressures of (final) consumers along the supply chains, which can be also associated to the concept of "water footprint of consumption" (see Hoekstra et al. 2009, 2011). From these references and many works which we will review, detailed accounting of the agri-food virtual water intensities along the supply chains were estimated, allowing to compute the water footprint of consumption at country level making use of the net trade of countries. In other words, having the (on site) water footprint of production of each country, the existing detailed trade data (from the Comtrade database of the United Nations and related databases) allows for computing "virtual water exports" and "virtual water imports". Adding the first one and subtracting the former one to the water footprint of production (i.e., adding the "net virtual water imports"), the water footprint of consumption of a country can be obtained (it seems clear then that it seems more difficult to obtain it in this way for a more specific region, for a particular type of consumption/household/person, etc.).

Input-output (IO) tables, and especially multi-regional input-output databases also aim for comprehensively capturing trade, integrating these with other sectoral transactions within the countries themselves. Particularly relevant in these type of databases[2], whose development and attention grew in recent years, has been the treatment of intermediate exports and imports, which are used for further processing,

[1]The water footprint was defined as an indicator of freshwater uses which examines both direct and indirect uses of water by a producer or consumer. Water footprints of an individual, business or community are defined as the total volumes of freshwater which are used to produce the goods and services consumed by them (an individual, business or community).

[2]WIOD (Dietzenbacher et al. 2013), EXIOBASE (Tukker et al. 2009, 2013; Wood et al. 2015), EORA (Lenzen et al. 2013; Aldaya et al. 2010), OECD (OECD 2016), GTAP (Narayanan et al. 2012, 2015), see also the Special Issue (Tukker and Dietzenbacher 2013).

re-export, etc. Also models associated to them have put particular focus on the rigorous accounting, in general with seminal works as Miller and Blair 2009, (Chenery 1953; Isard 1951; Moses 1955), and in more detail and with the focus of avoiding double counting when measuring those transactions and associated measures such as vertical specialization, value added in trade (VAiT, which looks at gross exports and imports), trade in value added (TiVA, with more of a footprints approach), etc. (Hummels et al. 1999; Dietzenbacher 2010; Stehrer 2012; Foster-McGregor and Stehrer 2013; Koopman et al. 2014). The discussions and chosen accounting methods hence have also importance on the way in which the embodied (virtual) water flows involved are computed. In any case, by placing the focus on the water footprints of consumption approach, a measure analogous to TiVA, i.e. the "trade in water" can be computed following the cited seminal works. As major drawback of these type of models, one usually finds the limitation of having to work at much more aggregate level of sectors/products than trade statistics allow for.

In this chapter we aim to put into perspective the pros and cons of input-output methods, and especially of multi-regional input-output models, for the estimates of water footprints, revealing possible aspects in which these are more capable of providing improvements, and viceversa.

The remainder of the chapter is organized as follows. The second section deals with a literature review focused on the methods and estimates of water footprint of consumer products. The third section focuses more on one of the approaches, which we consider particularly suited for the estimates of industrial products, which is the environmentally extended input-output tables and models. A few results obtained with these methods are discussed, while introducing and developing challenges in estimating water footprint of consumer products. The final discussion aims to sketch future lines of research on water footprint of consumer products, particularly the industrial ones, based on the identified challenges and opportunities.

2 Literature Review on Water Footprint of Consumer Products

Although as we will see in Sect. 3, even when concepts and estimates associated to embodied water contents existed since the late 60s, the concept of Virtual Water (VW) was defined in the 90s by (Allan 1993, 1994) as the amount of water that had to be used to generate a given product. Closely related to the concept of VW, the Water Footprint of a country (or region, business, etc.) is defined as the volume of water necessary for the production of goods and services which are consumed by the people thereof. Hoekstra and Hung (2002) and subsequent studies such as Hoekstra and Chapagain (2007) estimate the water footprint associated with the consumption of individuals, calculating the footprint of almost every nation in the world for the period 1997–2001. Moreover, in this and other subsequent studies, they distinguish between external and internal water footprint (domestic and foreign), and water colours (green, blue and grey). Then the international commodity trade flows involving water can be accounted for, following (Chapagain and Hoekstra 2004), departing from the national water resources, adding the net virtual water flows (imported-exported) which enters the country. Other studies calculating the global flows of virtual water are (De Fraiture et al. 2004; Oki et al. 2003; Zimmer and Renault 2003).

After those initial works, a large number of works were developed, which we do not aim to comprehensively review here. The Water footprint manual (in their different versions, (Hoekstra et al. 2009, 2011)) gather most of the more popular and cited concepts, methodologies, etc. Also some of the comprehensive reviews of water footprint studies are e.g. (Chenoweth et al. 2014) and (Zhang et al. 2017).

Despite the growing enthusiasm for the expansion and use of water footprints, several researchers also have raised concerns regarding the usefulness of the concept and estimates, highlighting the insufficient information provided on the opportunity cost of water, as an indicator of environmental impact and sustainability, and others (Wichelns 2010, 2011, 2015, Gawel and Bernsen 2011, 2013; Perry 2014). Again, we do not aim to comprehensively review these discussions here, and given that as indicated in many other works, (Hoekstra et al. 2009; Hoekstra et al. 2011; Vörösmarty et al. 2015), the concept may have limitations but does provide interesting insights on pressures as consumers and has been put in relation to economic productivity (Aldaya et al. 2010), water stress or scarcity (Chouchane et al. 2018; Makate et al. 2018; Lenzen et al. 2013), etc., we prefer to focus on the discussion of the methodologies and how to better get estimates on water footprints of consumer products. Furthermore, water footprints have often also stimulate discussions on interrelations between water use, food security and consumption (Cazcarro et al. 2012; Vanham and Bidoglio 2013; Hadjikakou et al. 2013; Vanham et al. 2013; Serrano et al. 2016), not only for the present, but also with historical perspectives (Duarte et al. 2014b). Furthermore, the concept has found increased attention at the corporate level (Ercin et al. 2011; Gerbens-Leenes and Hoekstra 2008; Mason 2013; Rivas Ibáñez et al. 2017).

In water footprints of specific crops, animal, food products and forest products may typically be better captured by the study of the specific supply chains of these products, as most of the cited works had done, given the great heterogeneity in water intensities among most of these categories. Similarly, the concept of water footprint was approached from frameworks such as life cycle assessment (LCA), (Milà i Canals et al. 2009; Pfister et al. 2009; Ridoutt and Pfister 2010; Yang et al. 2013; Pfister et al. 2017).

But also as reviewed in Daniels et al. (2011), Duarte and Yang (2011), Chenoweth et al. (2014), other studies have taken top-down approaches, as notably those works making use of input-output (IO) models. There is a large number of water IO studies (being these other with interests such as net trade, water footprint per capita, water use changes, etc. Hendricks and Dehaan (1981), Duarte (2002), Sánchez-Chóliz and Duarte (2005), Okadera et al. (2006), Velázquez (2006), Dietzenbacher et al. (2007), Guan and Hubacek (2007, 2008), Wang et al. (2009), Zhao et al. (2009), Yu et al. (2010), Zhao et al. (2010), Duchin and López-Morales (2012), López-Morales and

Duchin (2011, 2014), Feng et al. (2012), Duarte et al. (2014a), Zhang and Anadon (2014), Zhao (2014), Shi and Zhan (2015), Feng and Hubacek (2015), Arto et al. (2016), Serrano et al. (2016), Wang et al. (2016).

The question of bottom-up versus top-down approaches has been studied as well empirically in Feng et al. (2011). In the following section we focus on how water foot-printing has been developed making use of more top-down techniques and analyses, such as this extended environmental input-output (IO) models.

As we will try to show (in the results section) and discuss, even when this last type of models may suffer from important challenges for estimating water footprints, such as the aggregation issues which water intensities typically suffer due to limitations of data availability, we consider that they are relatively well suited for the estimate of WF consumer products, notably of industrial and even services ones, particularly for the computations of blue and grey water, which do not have such an agricultural focus as green water.

3 Methods for Environmental Input-Output Analyses

Following the final thread of literature reviewed, in this section we focus on one of the methods which has grown to estimate water footprints, notably to account for multi-regional flows, which is the extended environmental input-output models.

3.1 *Input-Output (IO) Models and Water*

Although already in 1968 we may find a first analysis of the uses of water using IO techniques (Lofting and McGauhey 1968), this methodology is not generalized in the following years and the analyzes on water resources are carried out basically with engineering approaches and with approaches of partial equilibrium, focusing on agricultural yields as well as on the study of regulatory processes (construction of wetlands and large canals). A large part of the studies developed during the first two thirds of the twentieth century follow the perspective of supply, for which the only problem was the possible availability of water from the courses of rivers, reservoirs and aquifer. The basic and simplistic idea was that the technological leap involved in irrigation justified the expenses and investments, if they had the financial capacity to carry them out.

This approach suffers a strong change in the last third period of the twentieth century, from a supply perspective to a demand one, motivated to a large extent by the appearance of strong limitations in the desired availability (irrigation techniques were very inefficient and the agrarian, industrial and service returns are very high), due to climate change and the strong social conflicts that develop in some places around water. This change of perspective makes it easier to connect with the IO models, since the Leontief model is basically a demand model. In addition, the

IO models allow a more integrated vision of all activities, agricultural, livestock, industrial and service, revealing their mutual interrelations or linkages.

Under this approach of demand and assuming the limitations of availability, analyzes have been carried out with IO models where the optimization of income or water productivity under restrictions was sought. Of these, the most common were the limitations of water availability, labour or the requirement of a minimum level of income. Frequently in these analyzes some type of technological change was introduced, associated with the search for efficiency in water uses. Works carried out basically in this framework, although some years later they are Vélazquez (2001) and Sánchez-Chóliz and Duarte (2000) that incorporate the technological change involved in the introduction of irrigation. Starting from Leontief's basic model,

$$\left.\begin{array}{l} x_1 = x_{11} + x_{12} + \ldots + x_{1n} + y_1 \\ x_2 = x_{21} + x_{22} + \ldots + x_{2n} + y_2 \\ \ldots\ldots\ldots\ldots\ldots \\ x_n = x_{n1} + x_{n2} + \ldots + x_{nn} + y_n \end{array}\right\} \Leftrightarrow \mathbf{x} = \mathbf{Ax} + \mathbf{y} \Leftrightarrow \mathbf{x} = (\mathbf{I} - \mathbf{A})^{-1}\mathbf{y}, \mathbf{L} = (\mathbf{I} - \mathbf{A})^{-1} \quad (1)$$

being x_i the gross output of good i ($\mathbf{x}$ is the vector of outputs of the economy); x_{ij} the intermediate consumption of good i by the sector or activity j to obtain its gross output x_j; y_i the final demand for goods i; $\mathbf{A} = (a_{ij})$ the matrix of technical coefficients defined as $a_{ij} = x_{ij}/x_j$; and the final demand and $\mathbf{L}$ the so-called Leontief inverse.

3.2 *Water Values, Embodiments of Water*

Starting from formalizations of the concept of labour value in multisectorial models such as that of Morishima (1973)(which uses a simple IO model), the existence of other values associated with other material inputs were considered (K-values for an input K of Vegara 1979, or the steel values of Roemer 1982). All of them identify these values with the direct and indirect requirements of the corresponding input, that is, with the amount of embodied input. Specifying these new values in the water input was a relatively simple jump that occurred in the early 90s (e.g. Sánchez Chóliz et al. 1995).

Being $\mathbf{w} = (w_i)$ a vector of unit coefficients of water consumption per unit of product and $\mathbf{A}$ the matrix of technical coefficients, we may obtain the water values, or embodied (or virtual) water per unit of final demand (representing the weighted sum of the backward linkage by the w_i):

$$\boldsymbol{\omega} = \mathbf{w}'(\mathbf{I} - \mathbf{A})^{-1} = \left(\sum_i w_i \alpha_{ij}\right) \text{verifying that } \mathbf{w}'\mathbf{x} = \mathbf{w}'(\mathbf{I} - \mathbf{A})^{-1}\mathbf{y} = \boldsymbol{\omega}'\mathbf{y} \quad (2)$$

As indicated in many virtual water and water footprint studies, without using a methodological base of IO, (Allan 1993, 1994, 1996) established the concept of virtual water, close to the one of embodied water (embedded). This concept will be reinforced later by the water footprint (Hoekstra and Hung 2002, showing the first definitions, and Chapagain and Hoekstra 2004, the first major computations). Initially these concepts were approached through trade relations between countries and with empirical estimates essentially bottom-up (from bottom to top, with process analysis), as opposed to top-type ones, typical of the IO demand models.

3.3 Multi-regional Input-Output Models and Water

Based on the same equations than above, but extending the definition not only to the intermediate transactions between sectors (i and j) but also to regions (r, s, etc.), we may compute the above (making use of multi-regional input-output databases) defining extended matrices. This type of model is grounded on the multiregional and interregional models of Isard (1951), Chenery (1953), Moses (1955), Miller and Blair (2009), being the equilibrium Eq. (3) as:

$$\mathbf{X}^* = \mathbf{A}^*\mathbf{X}^* + \mathbf{Y}^* \leftrightarrow \mathbf{X}^* = \left(\mathbf{I} - \mathbf{A}^*\right)^{-1}\mathbf{Y}^* = \mathbf{L}^*\mathbf{Y}^* \tag{3}$$

With

$$\mathbf{A^*} = \begin{bmatrix} \mathbf{A}_{11} & \mathbf{A}_{12} & \cdots & \mathbf{A}_{1n} \\ \mathbf{A}_{21} & \mathbf{A}_{22} & \cdots & \mathbf{A}_{2n} \\ \cdots & \cdots & \cdots & \cdots \\ \mathbf{A}_{n1} & \mathbf{A}_{n2} & \cdots & \mathbf{A}_{nn} \end{bmatrix}; \mathbf{X^*} = \left[\mathbf{x}_1, \mathbf{x}_2, \ldots, \mathbf{x}_n\right] = \begin{bmatrix} \mathbf{x}_{11} & \mathbf{x}_{12} & \cdots & \mathbf{x}_{1n} \\ \mathbf{x}_{21} & \mathbf{x}_{22} & \cdots & \mathbf{x}_{2n} \\ \cdots & \cdots & \cdots & \cdots \\ \mathbf{x}_{n1} & \mathbf{x}_{n2} & \cdots & \mathbf{x}_{nn} \end{bmatrix};$$

$$\mathbf{Y^*} = \left[\mathbf{y}_1, \mathbf{y}_2, \ldots, \mathbf{y}_n\right] = \begin{bmatrix} \mathbf{y}_{11} & \mathbf{y}_{12} & \cdots & \mathbf{y}_{1n} \\ \mathbf{y}_{21} & \mathbf{y}_{22} & \cdots & \mathbf{y}_{2n} \\ \cdots & \cdots & \cdots & \cdots \\ \mathbf{y}_{n1} & \mathbf{y}_{n2} & \cdots & \mathbf{y}_{nn} \end{bmatrix}$$

Being $\mathbf{A}^*$ the multi-regional technical coefficients matrix, $\mathbf{L}^*$ the so called Leontief inverse, being $\mathbf{X}^*$ the output matrix and $\mathbf{Y}^*$ the final demands matrix.

In $\mathbf{A}^*$, each $i \times j$ matrix $\mathbf{A}_{rr}$ specifies the domestic technical coefficients of region r. The $i \times j$ off-diagonal matrices $\mathbf{A}_{rs}$ designate the coefficients of region s of imported inputs originated in r. Each element a_{rs}^{ij} of matrix $\mathbf{A}^*$ states the quantity produced in r of output of sector i and consumed as input by region s in sector j per unit of total output in s of sector j.

Matrix $\mathbf{Y}^*$ is formed of n (being n the number of regions) column vectors $\mathbf{Y}_s$ with vector $\mathbf{Y}_{ss}$ ($i \times 1$) demonstrating the domestic final demand of s. $\mathbf{Y}_{rs}$ being the flows of finished products from r to s, or what is the same, final demands of goods from r to s which are not consumed as productive inputs.

$\mathbf{X}^*$ is formed of n column vectors $\mathbf{x}_s$ each of which according to (3) epitomises the production necessary to obtain final demand $\mathbf{y}_s$. Such production can be divided into $\mathbf{x}_{ss}$ (a $i \times 1$ vector), as showing production of s satisfying the final demand, and vectors quantifying the extra production which is necessary from other r regions.

We may get the water flows and also the water footprints, similarly to the green and blue virtual water flows and footprints (see Cazcarro et al. 2013). We can diagonalize these vectors to obtain:

$$\mathbf{Y}^{*+} = \begin{bmatrix} \widehat{\mathbf{y}}_{11} & \widehat{\mathbf{y}}_{12} & \cdots & \widehat{\mathbf{y}}_{1n} \\ \widehat{\mathbf{y}}_{21} & \widehat{\mathbf{y}}_{22} & \cdots & \widehat{\mathbf{y}}_{2n} \\ \cdots & \cdots & \cdots & \cdots \\ \widehat{\mathbf{y}}_{n1} & \widehat{\mathbf{y}}_{n2} & \cdots & \widehat{\mathbf{y}}_{nn} \end{bmatrix}$$

Making also use of $\widehat{\mathbf{w}}^*$, the matrix of the diagonalized water footprint of production per unit of economic output, we may get $\boldsymbol{\tau}^*$, which has dimension of (n*r) × (n*r), as also have each of the three elements shown in the following equation:

$$\boldsymbol{\tau}^* = \widehat{\mathbf{w}}^* \mathbf{L}^* \mathbf{Y}^{*+} \tag{4}$$

Estimating the double-entry matrix $\boldsymbol{\tau}_{\mathrm{r,s}}$ we indicate the virtual water flows from region r to s.

Each $\boldsymbol{\tau}_{\mathrm{r,s}}$ $(i \times j)$ is a matrix displaying from region r to region s the virtual water flows. The characterizing elements $\tau_{\mathrm{r,s}}^{i,j}$ symbolize the direct and indirect embodied water to satisfy the demands from region r *in* sectors i of region s of sector j. Then, with $\mathbf{e}$ being a column vector of ones, $\mathbf{e}'\boldsymbol{\tau}_{\mathrm{s,s}}\mathbf{e}$ is the water footprint generated in region s to sustenance its own final demand. In other words, it is the domestic component of the WF of region s, i.e., the internal water footprint of regional consumption of region s. Similarly, $\sum_{r \neq s}^{r} \mathbf{e}'\boldsymbol{\tau}_{\mathrm{r,s}}\mathbf{e}$ is the total virtual water import of region s (from all other regions), and $\sum_{s \neq r}^{s} \mathbf{e}'\boldsymbol{\tau}_{\mathrm{r,s}}\mathbf{e}$ the total virtual water export of region r (to all other regions). Hence, $\sum_{s \neq r}^{s} \mathbf{e}'\boldsymbol{\tau}_{\mathrm{r,s}}\mathbf{e} - \sum_{r \neq s}^{r} \mathbf{e}'\boldsymbol{\tau}_{\mathrm{r,s}}\mathbf{e}$ is the net export of virtual water. Moreover, $\sum_{r} \mathbf{e}'\boldsymbol{\tau}_{\mathrm{r,s}}\mathbf{e} = \mathbf{e}'\boldsymbol{\tau}_{\mathrm{s,s}}\mathbf{e} + \sum_{r \neq s}^{r} \mathbf{e}'\boldsymbol{\tau}_{\mathrm{r,s}}\mathbf{e}$ is the WF of consumption of region s.

3.4 Data Used for the Results and Challenges Discussion

In the following section we make use of the concepts of blue, green and grey water of the so cited Water Footprint approach (Mekonnen and Hoekstra 2011a)

As explained in Arto et al. (2012 and 2016), the World Input-Output Database (WIOD, see its description in Dietzenbacher et al. 2013) is made of a set of harmonized supply, use (SUT framework) and symmetric IO tables. These have been presented with valuations in previous year and current prices. It covered EU-27 countries plus other 13 major countries or regions in the world. Also the Rest of the World (RoW) was added as an aggregated region. The database also contains data on satellite accounts related to socio-economic and environmentally related indicators, such as emissions, energy, materials, land, etc.), but we focus on the water data. The years 1995–2009 is the time period considered. The accounts detail is 59 products, 35 industries as well as 5 final demand categories.[3]

Conventional national water uses accounts were limited to statistics on water withdrawals within their own nation. This considers essentially the use of surface and ground water by the various economic users and activities, being closer to the concept of blue water. The approach suggested by Mekonnen and Hoekstra (2010, 2011c) extends this framework by redefining blue water (consumed by products, different from the withdrawals) and includes as separate information the data on rainwater (green water use) and water volumes for load pollutants assimilation (grey water), to provide a wider view of humankind freshwater appropriation.

All in all, WIOD covers the uses of water (quantified in 1000 m^3) identifying three types: 1) Blue water, on the consumption of groundwater and surface water; 2) Green water, on the rainwater volume consumed, basically in crop production; and 3) Grey water, on the freshwater volume which is needed to assimilate wastes founded on ambient water quality standards.

In WIOD agricultural water use was estimated using crop and livestock water intensities from Mekonnen and Hoekstra (2010, 2011c) and livestock and crop production data from FAOSTAT. For the electricity sector the use of water for hydropower generation was obtained by means of the world average water use per unit of electricity (estimates of Mekonnen and Hoekstra 2011b) and the hydropower generation from the International Energy Agency (IEA). Water uses in other economic activities was obtained from the total water use in industry as shown in Mekonnen and Hoekstra (2011a), making use of the water use shares by industry from EXIOPOL as well as the gross output (constant prices) from WIOD. Households' water uses were estimated making use of the average domestic water supply (Mekonnen and Hoekstra 2011a) and population data from United Nations.

Although a new WIOD database (2016 release) has appeared for more recent years (2008–2014), the satellite accounts, and notably the water footprints of production ones have not been updated yet. The most recent year with data in the 2013 release was 2009, and so we have examined some of the figures looking at it.

[3]Available at the website (www.wiod.org) there is a download link and a detailed description of the database.

3.5 Extensions to Optimization Models, Water Restrictions, LCA-EIO Models, Etc

When we want to address the effects in an economy of greater or lesser availability of water, or the restrictions that suppose for example maximum levels of water pollution, we can make use of the presented framework (we do not make use here of $\oplus$ for the expansion for n regions) and extend it to solve an optimization problem with restrictions such as those posed in the dual problems that follow:

$$\left.\begin{array}{c}\text{Max } \mathbf{v}'x \\ x - Ax \geq y \\ \mathbf{w}'x \leq b \\ x \geq 0\end{array}\right\} \qquad \left.\begin{array}{c}\text{Min}[\, \boldsymbol{\gamma}' y + \mu b] \\ (\boldsymbol{\gamma}', \mu)\begin{pmatrix} I - A \\ w' \end{pmatrix} \geq v \\ \boldsymbol{\gamma} \leq 0 \\ \mu \geq 0\end{array}\right\} \tag{5}$$

If $\mathbf{v}$ is the vector of added value per unit of output and $\mathbf{w}$ the vector of water uses per unit of output, the first problem requires the maximization of the income or value added of the economy under the condition that a certain final and previously established demand is covered, and that the uses of water do not exceed a provision b. The second problem is equivalent to the first, but expressed in dual variables or shadow prices; $\boldsymbol{\gamma}$ gives the shadow prices of obtaining the fixed levels of demand (the variation in the optimum value added when the demand requirement increases by one unit) and μ the effect on the optimal added value of the increase of one unit in the initial allocation of water. Undoubtedly the problems to be addressed may be more complex, but their formulation will be essentially similar (see, for example, Sánchez-Chóliz and Duarte 2005, López-Morales and Duchin 2014). These type of applications try to minimize the use of water, but typically also of other factors of production, looking for the global costs minimization (see Duchin 2005; Cazcarro et al. 2016b). In the final discussion we will refere back to these models to comment on future and open lines of research which can improve the interest of water footprint studies for policy, scenarios, etc.

We have also referred to process analysis type of studies, LCA, LCI (Life Cycle Inventory), etc. as examples of bottom-up approaches, and to input-output models as an example of top-down, and it may seem that they are completely different frameworks. On the contrary, following intitial attempts of adding input-output results which were based on disaggregation of monetary input-output tables or also process based models (Bullard and Pilati 1976), a highly relevant literature has rigourously addressed their similarities (Suh 2004a, b; Suh et al. 2004; Suh and Kagawa 2005; Suh and Heijungs 2007; Ferrão and Nhambiu 2009), and formulated the hybrid LCI-IO framework or also so called Hybrid EIO-LCA (process-based LCA and economic input-output analysis-based LCA). The Waste Input-Output model (Kondo and Nakamura 2003; Nakamura 1999; Nakamura and Kondo 2002, also a generalization of the Leontief–Duchin environmental input–output (EIO) model Duchin 1990; Leontief 1970) can be also seen as a general framework for hybrid LCA involving waste management and recycling, and hence it can be of particular interest for modelling waste water treatment.

In LCA, a usual algebraic formulation is to consider an external demand of process output i be given as k. The use of tilde indicates any activity in the foreground system. When the technical coefficients of the foreground system quantify the commodities or products which are needed in each process, for achieving one unit activity level, t, then the technical coefficients are symbolized by, $\widetilde{\mathbf{A}}$ and:

$$\widetilde{\mathbf{A}} \cdot \mathbf{t} = \mathbf{k} \tag{6}$$

Taking into account that the environmental burdens associated with the processes in the foreground system are denoted by $\widetilde{\mathbf{w}}$ (representing water in this case), the environmental considerations are identified in the foreground process as:

$$\mathbf{W} = \widetilde{\mathbf{w}} \cdot \widetilde{\mathbf{A}}^{-1} \cdot \mathbf{k} \tag{7}$$

Matrix $\widetilde{\mathbf{w}}$ is the *intervention matrix*, given that its coefficients show interventions of the various economic processes in the environment: inputs (mainly extractions of resources, in this case water) and outputs.

Taking into account this formulation of the emissions for the foreground system (7) together with the one for IO above (in which $\mathbf{w}$, the vector of water uses per unit of output, is the background environmental matrix here) the Hybrid method can be expressed by the general expression below:

$$\mathbf{W} = [\langle \widetilde{\mathbf{w}} | \mathbf{w} \rangle] \cdot \begin{bmatrix} \widetilde{\mathbf{A}} & \mathbf{M} \\ \mathbf{N} & \mathbf{I} - \mathbf{A} \end{bmatrix}^{-1} \cdot \mathbf{k} \tag{8}$$

Following (Suh and Huppes 2005) the methodology to create the matrix of coefficients and to normalize the foreground and background units of the process, follows (9) and (10). **M** and **N** signify respectively inputs from foreground and background systems to one another. The dimension of elements for **M** and **N** should match with corresponding columns and rows. **M** displays total physical output per total production in monetary term, while **N** displays monetary input to each sector per given operation time. Being $\mathrm{q_{pq}}$ the input of sector p in each unit process q; $\mathrm{p_p}$ the unit price of product from sector p; and being $\mathrm{a_{pq}}$ the economic matrix technical coefficient.

$$\mathrm{l_{pq}} = \mathrm{q_{pq}} \times \mathrm{p_p} \tag{9}$$

$$\mathrm{m_{pq}} = -\mathrm{a_{pq}} / \mathrm{p_p} \tag{10}$$

4 Results and Challenges in Estimating Water Footprint of Consumer Products

As cited above, a number of articles have dealt with different tools but also challenges in water footprint estimates (e.g. among others, Hoekstra et al. no date; Younger 2006; Schendel et al. 2007; Lopez-Gunn and Llamas 2008; Gerbens-Leenes et al. 2009a, b; Hoekstra and Mekonnen 2012; Vanham et al. 2013; Aivazidou et al. 2016; Pellicer-Martínez and Martínez-Paz 2016; Pfister et al. 2017). Here we focus on those aspects, but also challenges which concern more the estimates of water footprints of consumption, i.e. of how to compute water footprints in consumer products, highlighting key interrelations and sectors involved.

4.1 *From Agricultural Centered Water Footprints, to Further Role for Industrial Ones*

A quick review of the water footprint studies reveals that most of the studies have been centered in agriculture, and then in the downstream supply chains derived from crop and animal products. As shown in Fig. 1 (Table A1 provides the codes of the regions) with the cited WIOD data and satellite accounts, undoubtedly around 60% or 70% of water footprints of production come from green water, which is estimated in most works only for agriculture and those derived products.

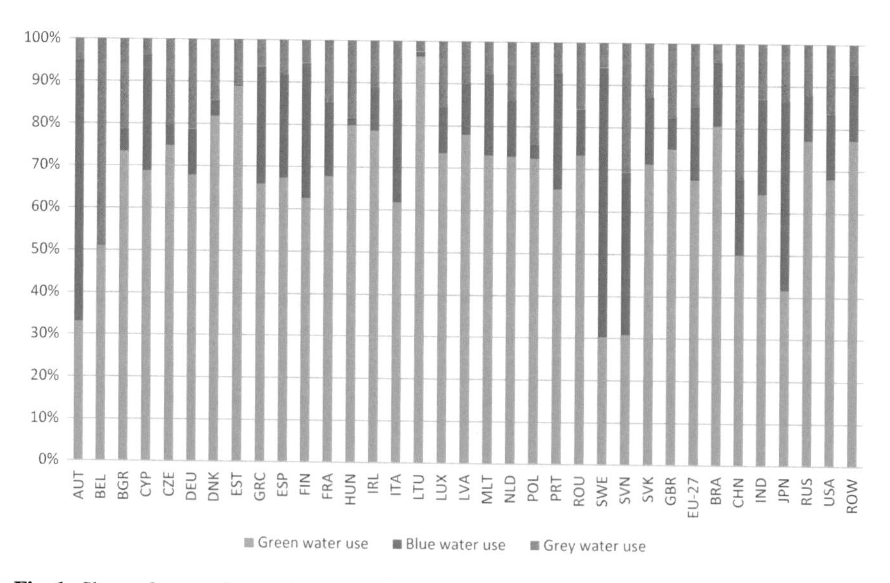

Fig. 1 Share of types of water by region (year 2009). *Source* Own elaboration from WIOD data (Release 2013) and environmental accounts

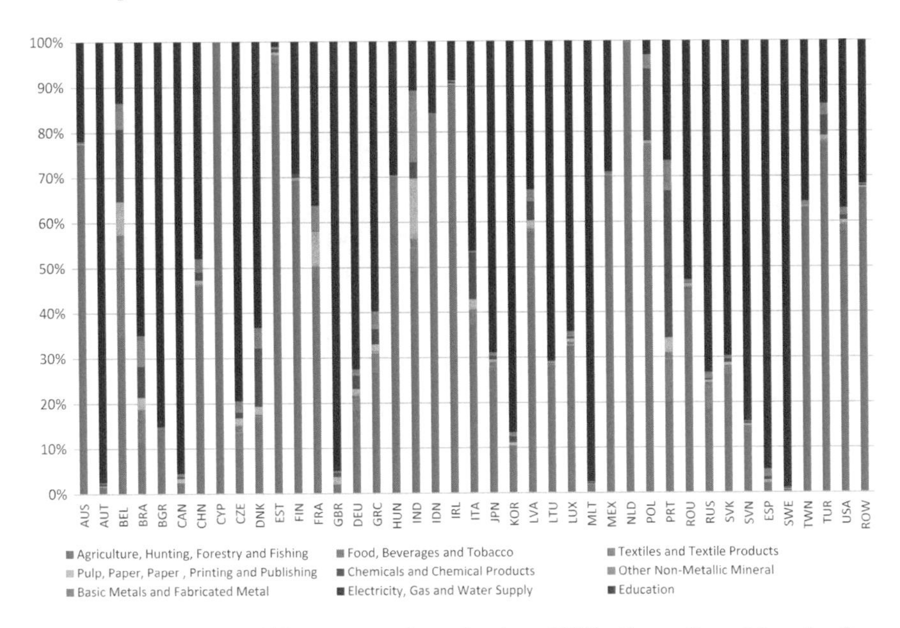

Fig. 2 Share by sectors of blue water use by region (year 2009). *Source* Own elaboration from WIOD data (Release 2013) and environmental accounts

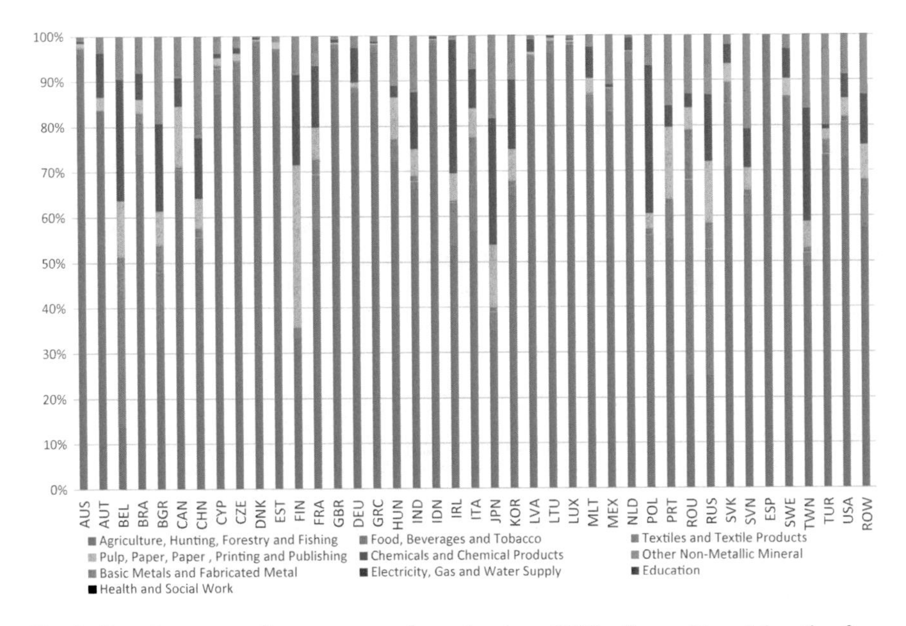

Fig. 3 Share by sectors of grey water use by region (year 2009). *Source* Own elaboration from WIOD data (Release 2013) and environmental accounts

As explained in Hoekstra (2013), water is not only used in the domestic context, or in agriculture, but also in industry in the production of commercial goods, such as processed food, paper, textiles, etc. Making use of a number of case studies, it was illustrated how water use throughout supply chains. It was also shown how water consumption is often linking distant places performing water uses and consumptions of goods and services. As an example, it was calculated that 15,000 litres of water are needed to produce 1 kg of beef, or 8,000 litres of water to produce a pair of jeans. Similar works have been refining these estimates, and analyzed the sustainability of consumer products, e.g. (Chico et al. 2013). Another line of research importantly related with other industrial sectors is found regarding the estimates of water footprint of bioenergy (Gerbens-Leenes et al. 2009a, b).

In the case of blue water, apart from the about 56% of water used in agriculture globally (similarly about 58% for the EU-27), also about 40% comes from the water use made by the sector of "Electricity, Gas and Water Supply". In terms of intensity (water per unit of sectoral output), the two sectors also stand out in the top, being the highest one of all that of the "Electricity, Gas and Water Supply" of Canada, with 2,317 litre/US$ (1000 i.e. m^3/ millions of US$ or m^3/1000 US$), far away from the 2nd one, of the same sector in Brazil, with 1,122 litre/US$; the third one is the agricultural sector of India, with 1,007 litre/US$ (Fig. 2).

In the case of grey water, about 60% of grey water is accounted for agriculture globally (similarly about 64% for the EU-27), also about 12% is grey water footprint of production in the sector of "Chemicals and Chemical Products". Globally also about 13% is grey water footprint of production in the sector of "Basic Metals and Fabricated Metals". In Europe, also notably 10% is grey water footprint of production in "Food, Beverages and Tobacco" sectors (globally it is 6%). Finally, also the sector of "Pulp, Paper, Paper, Printing and Publishing" typically accounts for about 5–6% of the grey water footprint (Fig. 3).

Furthermore, in terms of direct coefficients, we may see in Table 1 how although still the agricultural sector dominates in the first 20 highest intensities of grey water footprint of production, other sectors also imply large direct coefficients, being the first the "Textiles and Textile Products" sector of Russia, with 571 litre/US$ (1000 i.e. m^3/ millions of US$ or m^3/1000 US$),

Table 1 Intensities (direct coefficients) of grey water footprint of production in the year 2009 (litre/US$). *Source* Own elaboration from WIOD data (Release 2013) and environmental accounts

Rank	Intensity	Sector	Region
1	571	Textiles and Textile Products	RUS
2	523	Pulp, Paper, Paper, Printing and Publishing	IND
3	518	Chemicals and Chemical Products	BGR
4	445	Pulp, Paper, Paper, Printing and Publishing	RUS
5	422	Agriculture, Hunting, Forestry and Fishing	IND

4.2 *Specific (Industrial) Sectors/Products from with Important Water Footprints of Consumption are Obtained*

We have seen that the water footprints of production still reveal a very important role of agriculture, with a few other key (mostly industrial) sectors with secondary importance, but relevant for blue and grey water footprints. When we look at sectors/products with important embodied (virtual) water, leading to compute water footprints of consumption differently for regions, we may see that all sectors/products of the economy (35 in this case from WIOD) are involved.

Again, the numbers for green water guide the largest volumes, and hence we may see that the agricultural sector still is the one that provides more volumes of water to final consumers, with about 4,000 km^3 globally, followed by the food, beverages and tobacco sector, with about 2,300 km^3 globally. At a larger distance, sectors such as Hotels and restaurants provide to final consumers about 400 km^3 globally, but also a clear industrial one, such as the construction sector, provides more than 300 km^3.

When we look, as we do in Fig. 4, at blue and grey water, the role of agriculture and the agri-food supply chain reduces its role. In this case we rank the 20 sectors/products having the highest total (blue, green and grey) embodied (virtual) water for final uses. In other words, we look at the intensities in the consumption categories. In this case the construction sector appears even with more importance (implying that the long supply chains of the construction sector which involve high resource use, also implies relevant volumes of blue water uses, and of water pollution).

In the case of China, as we show in Fig. 5, according to the estimates with WIOD the blue water footprint of consumption comes more from the construction sector

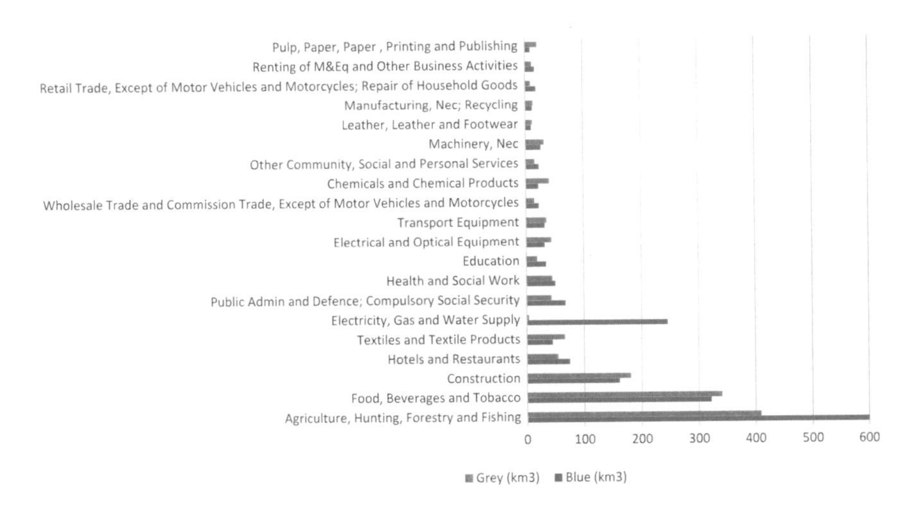

Fig. 4 Blue and grey water footprints of consumption by sector/product of final destination (km^3). *Source* Own elaboration from WIOD data (Release 2013) and environmental accounts

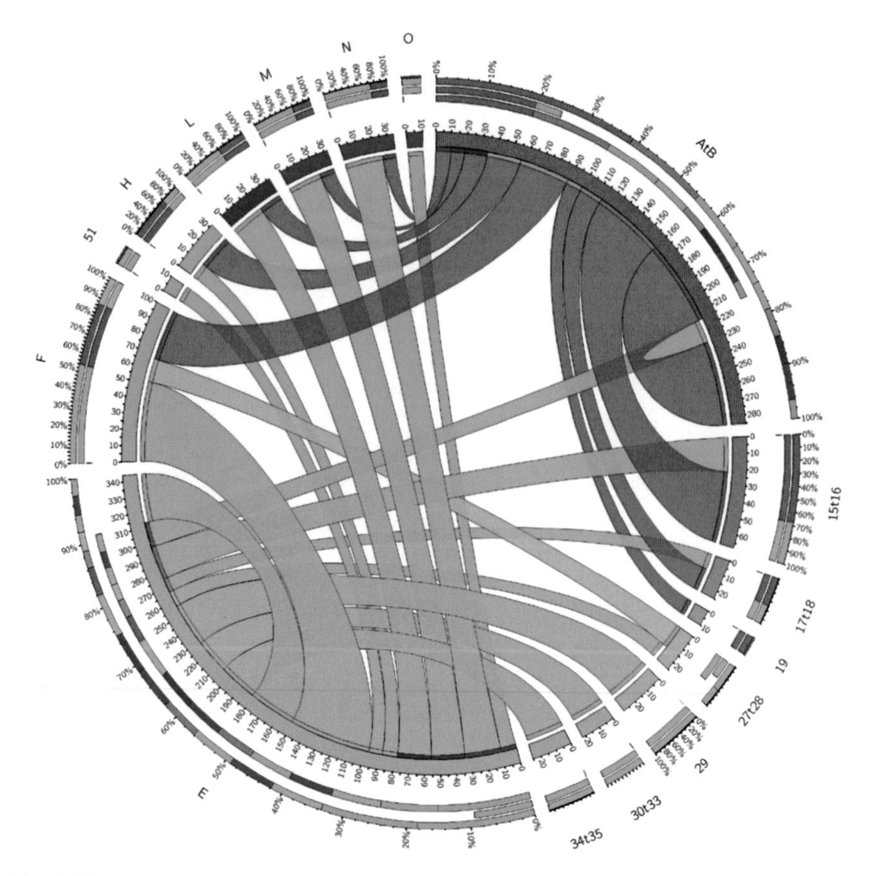

Fig. 5 Blue water consumption in China embodied (virtual) in final demand in China. Notes (for Figs. 5 and 6) on sectors: AtB: Agriculture, Hunting, Forestry and Fishing; C: Mining and Quarrying; 15t16: Food, Beverages and Tobacco; 17t18: Textiles and Textile Products; 19: Leather, Leather and Footwear; 20: Wood and Products of Wood and Cork; 21t22: Pulp, Paper, Paper, Printing and Publishing; 23: Coke, Refined Petroleum and Nuclear Fuel; 24: Chemicals and Chemical Products; 25: Rubber and Plastics; 26: Other Non-Metallic Mineral; 27t28: Basic Metals and Fabricated Metal; 29: Machinery, Nec; 30t33: Electrical and Optical Equipment; 34t35: Transport Equipment; 36t37: Manufacturing, Nec; Recycling; E: Electricity, Gas and Water Supply; F: Construction; 50: Sale, Maintenance and Repair of Motor Vehicles and Motorcycles; Retail Sale of Fuel; 51: Wholesale Trade and Commission Trade, Except of Motor Vehicles and Motorcycles; 52: Retail Trade, Except of Motor Vehicles and Motorcycles; Repair of Household Goods; H: Hotels and Restaurants; 60: Inland Transport; 61: Water Transport; 62: Air Transport; 63: Other Supporting and Auxiliary Transport Activities; Activities of Travel Agencies; 64: Post and Telecommunications; J: Financial Intermediation; 70: Real Estate Activities; 71t74: Renting of M&Eq and Other Business Activities; L: Public Admin and Defence; Compulsory Social Security; M: Education; N: Health and Social Work; O: Other Community, Social and Personal Services; P: Private Households with Employed Persons. Notes (for Figs. 5 and 6) on representation: The colour of the flow is given by the origin sector/product (e.g. in red from AtB; in green from E). Cells below 50% percentile are not shown; cells with small values are attenuated. *Source* Own elaboration from WIOD data (Release 2013) and satellite accounts, using the Circos software developed by Krzywinski et al. (2009)

(67 km^3) than from Agriculture, Hunting, Forestry and Fishing (60 km^3) or from Food, Beverages and Tobacco (40 km^3). Globally (Fig. 4) also Textiles and textile products reveal a predominant role, especially in the need of polluting water (as reflected in the grey water volumes, i.e. the water needed for dilution or assimilation of the load of pollutants given the context and standards). Interestingly, also some of the services sectors (not exactly "products" of consumption, but certainly economic activities purchased by final consumers, and involving water footprints along the supply chain) appear in this ranking, even before some other industrial activities. Obviously the large size of the final demands (and ultimately of the economies) globally, and especially in developed/most favoured countries (typically with economies accounting for 60–70% of GDP from services) mostly drives the results. In other words, truly these are not the most water (footprint of consumption) intense activities, but it is revealed that they cannot be totally excluded from the analysis of virtual water along the supply chains (as it has occurred in some agri-food focused studies, using process analysis/LCA, etc.). Notably the largest blue water footprints of consumption come from the "Electricity, Gas and Water Supply" in countries such as Sweden, Canada, Austria, Slovenia, Japan, Latvia, Brazil and several others.

Complementing the comments above, we may observe in Figs. 5 and 6 a case for China of water flows across sectors. In particular, having computed all the blue water flows with the multi-regional input-output model for all WIOD countries, we highlight in Fig. 5 blue water footprint of production from China which ends up in the final consumption of Chinese consumers according to those estimates. In Fig. 6 we register the same but per monetary unit of final demand (in this case given the homogeneous treatment of monetary units in WIOD it is litre/US$ (the original data is in 1000 m3/millions of US$ = m3/1000 US$ = litre/US$), but we are more concerned with the relative volumes across sectors); in other words, we account for the embodied (virtual) blue water intensities per unit of final demand. As we have seen through the chapter, as expected the main departing flows (Fig. 5) and main high intensities (Fig. 6) correspond to the sectors/products "AtB: Agriculture, Hunting, Forestry and Fishing" (in red) and "E: Electricity, Gas and Water Supply" (in green). As explained above and now shown in Fig. 5, a main recipient of these flows (from those activities, and "Basic Metals and Fabricated Metal") is the (sectors/products) consumption category "F: Construction", becoming the main sector/product through which final consumers in China get their water footprint of consumption. One may observe this by focusing only on the parts of the ribbon which just have incoming flows. In the case of "F" all represented flows (cut-off of 50% highest) are incoming; while in the case e.g. of "AtB" only the last two on the right hand side (from numbers 220–290, which does not have an equivalent to a figure in km^3) are inflows from "E" and especially from the sector itself (55 km^3 of water footprint of "AtB" which ends up in "AtB" sold to final Chinese consumers). Also in the case of "E", the only shown inflow is coming from the sector itself. We may also notice how the sectors/products of "15t16: Food, Beverages and Tobacco" embody non-marginal volumes in final demand (which came originally from the former two cited sectors/products) but for blue water in China it is not as large as "F". In the case of the intensities (Fig. 6), we also see that from this sector the water intensities embodied (virtual) in final demand

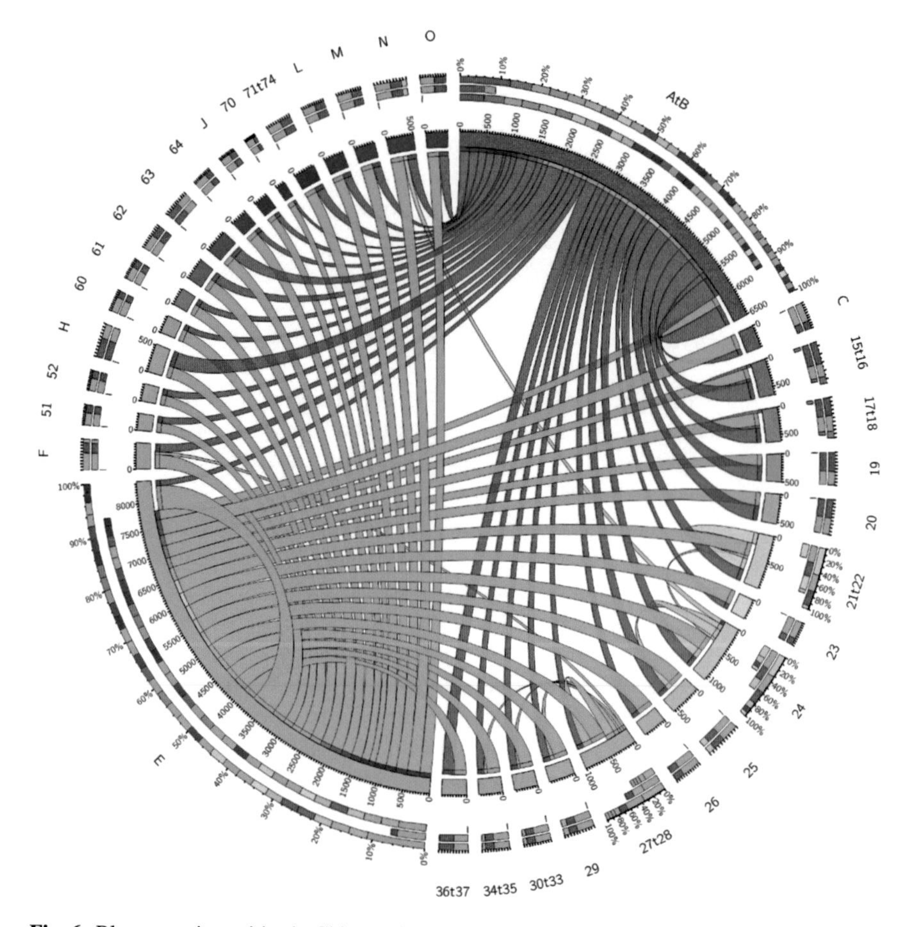

Fig. 6 Blue water intensities in China embodied (virtual) in final demand in China. *Source* Own elaboration from WIOD data (Release 2013) and satellite accounts, using the Circos software developed by Krzywinski et al. (2009)

are not so much larger than of many other industrial sectors. Having the caution of considering that this is likely to be somehow an artifact or a bias of input-output models, these intensities still reveal though that the sector has the highest intensity of water originated in agriculture (red flow), followed by "H: Hotels and Restaurants"; "19: Leather, Leather and Footwear"; "20: Wood and Products of Wood and Cork" and "17t18: Textiles and Textile Products", which are also consumer products with high embodied (virtual) water intensities per unit of final demand.

Furthermore, most of these cited (products) consumption categories and especially a few other industrial ones also have important embodied (virtual) water intensities due to inputs from "E: Electricity, Gas and Water Supply". In particular this is clear in the cases of "26: Other Non-Metallic Mineral", "27t28: Basic Metals and Fabricated Metal", "C: Mining and Quarrying", "24: Chemicals and Chemical Products";

"25: Rubber and Plastics"; "23: Coke, Refined Petroleum and Nuclear Fuel", "F: Construction" and "21t22: Pulp, Paper, Paper, Printing and Publishing". In the case of this last sector, it also stands out the high embodied (virtual) water intensities due to water footprint of production of the sector itself. Following with some of the insights obtained when examining the water footprints of production, apart from the cited "AtB", "E" and "21t22", only "27t28: Basic Metals and Fabricated Metal" and "24: Chemicals and Chemical Products" have some relevant water intensities, which in this case are revealed in the presence of some flows within the sector, meaning water embodied (virtual) in final demand in China from the water consumption from the sectors themselves.

The estimates presented above, compared to water process analysis type of studies, in which (only) very specific and detailed agri-food chains are studied, have a major advantage, and a major drawback. The advantage is a full accounting of all the interactions among economic sectors, notably industrial and services ones, which allows capturing these economic interrelations, which also involve embodied (virtual) water transfers (see e.g. Lenzen 2009 for rationale on this). The major drawback is that water is most present precisely in those agri-food chains, which hence in a multi-regional model as the one presented is considered in a very rough way. In particular, the water intensities of the whole agricultural sector are treated homogeneously (when in reality they are quite heterogeneous) and hence the same intensities are considered in agricultural and forestry goods sold for transformation (to the food industry, textiles, paper industry, trade sector, etc.) no matter whether they are cereals, vegetables, meat or wood. Clearly this major drawback is more evident for the green water estimates.

We consider that an interesting line of research hast do with providing indications on how these estimates can be improved for both types of studies. Furthermore, this is not the only great challenge in the world of water footprints estimates. Also when integrations of methods having been attempted other discrepancies, biases, errors, etc. have been found. For example, since the idea of this chapter was to show light on the general literature, methods, trends, etc. of water footprints of consumer products, placing an important focus on the industrial interactions and key sectors involved, we have not emphasized the global numbers of water footprints. In this regard, in the data we have seen from WIOD, from 1995 to 2009 the global water footprints ranged annually from 8,700 km^3 to 12,000 km^3 (as we have shown, representing about 70% green, 17% blue and 13% grey). This means that in the year 2007, having a total of 11,445 km^3, close to 8,000 km^3 was green water and 1,900 km^3 of blue water footprint of production. If we look at the satellite accounts of EXIOBASE version 2.2. Tukker et al. (2009), we find that about 4,480 km^3 is green, while about 1,600 is blue water footprint of production. In the case of green water, we may see that in EXIOBASE 4,423 km^3 correspond to crops, and 0,039 km^3 to animals. This clearly contrasts with most of water footprint studies, such as Mekonnen and Hoekstra (2010, 2011c), where already the water footprint of crops represents about 5,000 km^3, and the water footprint of animal products and animals represent more than 2,100 km^3. Certainly there should be a recognition of the framework in which only the direct water consumption by animals is taken into account (while water footprint from feed comes from the purchases of those animal categories to the crops categories, from the processed feed of the food industry, etc.), however, obviously the discrepancies stand out. In the case of WIOD data, we will see that also some error may have been computed by double counting some of the water from feed for animals which is not fodder, but from crops (already accounted in the water footprint of crops), but definitely this represents a much smaller bias/error. In EORA, for this year the global volumes provided in the satellite accounts (which are the same yearly for all the period 1990–2013), is estimated with a total of 7,601 km^3, being 341 km^3 grey water and the rest estimates of blue and green water footprint of production from crops and all the other sectors.

Interestingly then, typically discrepancies in the initial data taken, as direct uses or consumption of water, as it also happens for the accounting of other resources, tend to have importance. On the contrary, with the results of water footprints of consumption, the characterization of sectors/products involved usually match quite well. Indeed, for the studies of consumption-based carbon footprints, it has been found important aspects of convergence between the widely used multi-regional input-output databases, such as EORA, EXIOBASE, WIOD and OPENEU's (Moran and Wood 2014).

5 Conclusions and Final Discussion

5.1 Conclusions

With current general estimates of grey water footprints, and notably blue water footprints of production do have a less relevant role of agriculture (still around 60%) than green water footprints. When we look at the water footprints of consumption, the numbers for green water guide the largest volumes, and hence we may see that the agricultural and food industry sectors still are the ones that provides more volumes of water to final consumers. When we look at blue and grey water, the role of agriculture and the agri-food supply chain reduces its role, and the role of Electricity, gas and water Supply increases, especially in the case of blue water footprints. This insight provides also foundation to further look into the details of the water footprints in the water and power sectors. Sectors/products such as construction, textiles and even services appear as relevant. Obviously the large size of the final demands globally, and especially in developed/most favoured countries mostly drives the results, but it is revealed their importance when considered fully the supply chains. As we discuss below, this has important implications for the consideration of different methodologies for different purposes, or ideally, integration of best capabilities and insights of each "worlds/methods", including among these other not so frequently used or fully integrated, as other hydrological models, water scarcity indicators, water management constraints, scenarios for the future, etc.

5.2 Further Discussion

Based on a large thread of literature, but also from the evidence of this chapter, the integration of frameworks, notably of bottom-up or process based analysis, and top-down (particularly multi-regional input-output models) seems particularly desirable for the estimates of water footprints. This can go both ways (or even in three, if we consider integrations such as the LCA-EIO models presented in Sect. 3.5), i.e., on the one hand, by extending the supply chains computed in process based analyses, by incrementally integrating extending the boundaries considered according to the main sectors/products identified with the full accounting of supply chains. On the other hand, by using the process based information, of LCA, of specific supply chains accounting, etc. to disaggregate multi-regional input-output databases. Lenzen (2011) argued that disaggregation of IO data, even if it would be grounded on a small number of real data points, is revealed a preferred option to aggregating environmental data, in order to get IO multipliers. When disaggregated sectors are heterogeneous in comparison to their environmental and economic features, this becomes even more of an evidence. At this point, it should have been made clear the importance of going down into detailing the agri-food sectors in the MRIOs, if possible (if data availability and quality allows for it) reaching to the product level. In this regard, it is worth connecting with the insights learned from those other methodologies which are more focused on specific products and processes. For example, the case studies examined by Ridoutt et al. (2009) of water footprinting at the product brand level revealed several issues, namely the lack of correspondence between water footprints and the water availability for alternative uses when no production occurs, as well as the challenge in relating water footprints to potential social and environmental harm. As such, it was concluded that the water footprinting concept needs further development to be useful for promoting sustainable production and consumption.

We have indicated that the question of bottom-up vs. top-down approaches has been studied as well empirically in Feng et al. (2011). The differences they found in the WEBT (Water Embodied in Bilateral Trade) and MRIO (Multi-regional Input–Output Analysis) approaches were due to an inter-regional cut-off effect. While the MRIO approach traces entire global supply chains, the WEBT method only traces domestic supply chains. They also found that both top-down and bottom-up approaches are highly dependent on the quality of datasets already existent, being quite different between each other. The total water footprints of countries on the basis of several approaches varied by up to 48%, and this difference was even larger when looking at sectoral level. MRIO (Multi-regional Input–Output Analysis). This single results should be enough to still consider the importance of MRIO in an increasingly interconnected and globalized world. We add that this is of importance for the estimate of WF consumer products, notably of industrial and even services ones, particularly for the computations of blue and grey water, which do not have such an agricultural focus as green water.

5.3 *Open and Future Lines of Research*

According to the above, it seems quite clear that one of the main lines of development in the theme of water is wider integrated analyses. This goes for the highlighted question of methodologies, but can also be extended to connecting with many of the concerns and critiques, related to the questions of comparative advantage (i.e., looking at water but also at other factors of production in an integrated manner), to indicators which better help into water management and reflect scarcity, etc. In this regard, formulations such as those presented in Sect. 3.5 may allow increasing the interest of water footprint studies for policy. For example, optimization models can be run (minimization can be implemented of water footprint of production or water uses, but also of water costs, of global resources uses or global costs, etc.) to satisfy current (or future, under certain scenarios, notably with uncertainties from climate change, etc.) water footprints of consumption, revealing the comparative advantage of each of the regions by specializing in the production of different sectors/products.

We highlight the desirable increased integration meaning also looking at the problems in a wider geographic context and if possible globally of the entire planet. Multi-regional models are making this task easier and easier, allowing integration of economic activity and impacts in very different places. This global analysis should also connect consistently with the relevant scale of the study areas (e.g. river basins). For example, the impacts can be addressed in particularly sensitive areas, arid, with scarcity, s, etc., linking models and results with other hydrological, agrarian, Geographic Information Systems (GIS), etc. It should also contribute with guidelines for the design of new water policies that will face the possible effects of climate change and of large urban concentrations.

Given also some of the critiques on the further development to be of use for encouraging sustainable production and consumption (e.g. Ridoutt et al. 2009; Perry 2014), it seems interesting to also link water footprints of consumer products back to production (see Hubacek et al. 2016; Cazcarro et al. 2016a), and to water scarcity indicators (e.g. Brown and Matlock 2011) and relative measures so that they can be helpful indicators for water resource management.

Appendix

See Table A1.

Table A1 Codes and names of regions in WIOD used in this chapter

AUS	Australia	ITA	Italy
AUT	Austria	JPN	Japan
BEL	Belgium	KOR	South Korea
BGR	Bulgaria	LTU	Lithuania
BRA	Brazil	LUX	Luxembourg
CAN	Canada	LVA	Latvia
CHN	China including Hong Kong	MEX	Mexico
CYP	Cyprus	MLT	Malta
CZE	Czech Republic	NLD	Netherlands
DNK	Denmark	POL	Poland
ESP	Spain	PRT	Portugal
EST	Estonia	ROU	Romania
FIN	Finland	RoW	Rest of the world
FRA	France	RUS	Russia
GBR	United Kingdom	SVK	Slovenia
DEU	Germany (until 1990 former territory of the FRG)	SVN	Slovakia
GRC	Greece	SWE	Sweden
HUN	Hungary	TWN	Taiwan
IDN	Indonesia	TUR	Turkey
IND	India	USA	United States
IRL	Ireland	EU-27	European Union of the 27 countries

References

Aivazidou, E., et al. (2016). The emerging role of water footprint in supply chain management: A critical literature synthesis and a hierarchical decision-making framework. *Journal of Cleaner Production Elsevier Ltd, 137,* 1018–1037. https://doi.org/10.1016/j.jclepro.2016.07.210.

Aldaya, M. M., Martínez-Santos, P., & Llamas, M. R. (2010). Incorporating the water footprint and virtual water into policy: Reflections from the Mancha Occidental Region, Spain. *Water Resources Management, 24*(5), 941–958. https://doi.org/10.1007/s11269-009-9480-8.

Allan, J. A. (1993) Fortunately there are substitutes for water otherwise our hydro-political futures would be impossible. In *Priorities for Water Resources Allocation and Management* (pp. 13–26).

Allan, J. A. (1994) Overall perspectives on countries and regions. In *Water in the Arab World: Perspectives and Prognoses* (pp. 65–100).

Allan, J. A. (1996) Policy responses to the closure of water resources: Regional and global issues. In P. Howsam, & R. C. Carter (Eds.), *Water Policy: Allocation and Management in Practice. London Proceedings of International Conference on Water Policy*. Cranfield University.

Arto, I., et al. (2012) *Global resources use and pollution, volume 1/ production, consumption and trade (1995–2008).* The JRC Institute for Prospective Technological Studies (JRC-IPTS). https://doi.org/10.2791/94365.

Arto, I., Andreoni, V., & Rueda-Cantuche, J. M. (2016). Global use of water resources: A multiregional analysis of water use, water footprint and water trade balance. *Water Resources and Economics, 15,* 1–14. https://doi.org/10.1016/j.wre.2016.04.002.
Brown, A., & Matlock, M. D. (2011). *A review of water scarcity indices and methodologies*.
Bullard, C. W., & Pilati, D. A. (1976). *Reducing uncertainty in energy analysis*. Urbana, IL: University of Illinois.
Cazcarro, I., Duarte, R., & Sánchez-Chóliz, J. (2012). Water flows in the Spanish economy: Agri-food sectors, trade and households diets in an input-output framework. *Environmental Science and Technology*, *46*(12). https://doi.org/10.1021/es203772v.
Cazcarro, I., Duarte, R. & Sánchez Chóliz, J. (2013). A multiregional Input-Output model for the evaluation of Spanish water flows. *Environmental Science and Technology*, *47*(21), 12275–12283. https://doi.org/10.1021/es4019964.
Cazcarro, I., Duarte, R., & Sánchez Chóliz, J. (2016a). Tracking water footprints at the micro and meso scale: An application to Spanish tourism by regions and municipalities. *Journal of Industrial Ecology*. https://doi.org/10.1111/jiec.12414.
Cazcarro, I., López-Morales, C. A., & Duchin, F. (2016b). The global economic costs of the need to treat polluted water. *Economic Systems Research*, *28*(3). https://doi.org/10.1080/09535314.2016.1161600.
Chapagain, A. K., & Hoekstra, A. Y. (2004). *Water footprints of nations, value of water. Research report series*.
Chenery, H. B. (1953). Regional Analysis. In H. B. Chenery, P. G. Clark, & V. C. Pinna (Eds.), *The structure and growth of the Italian economy* (pp. 98–139). Rome: U.S. Mutual Security Agency.
Chenoweth, J., Hadjikakou, M., & Zoumides, C. (2014). Quantifying the human impact on water resources: A critical review of the water footprint concept. *Hydrology and Earth System Sciences, 18*(6), 2325–2342. https://doi.org/10.5194/hess-18-2325-2014. Copernicus Publications.
Chico, D., Aldaya, M. M., & Garrido, A. (2013). A water footprint assessment of a pair of jeans: The influence of agricultural policies on the sustainability of consumer products. *Journal of Cleaner Production, 57,* 238–248. https://doi.org/10.1016/j.jclepro.2013.06.001. Elsevier Ltd.
Chouchane, H., Krol, M. S., & Hoekstra, A. Y. (2018). Virtual water trade patterns in relation to environmental and socioeconomic factors: A case study for Tunisia. *Science of the Total Environment, 613–614,* 287–297. https://doi.org/10.1016/j.scitotenv.2017.09.032.
Daniels, P. L., Lenzen, M., & Kenway, S. J. (2011). The ins and outs of water use–A review of multi-region input–output analysis and water footprints for regional sustainability analysis and policy. *Economic Systems Research, 23*(4), 353–370. https://doi.org/10.1080/09535314.2011.633500. Routledge.
Dietzenbacher, E., et al. (2007). Analysing Andalusian virtual water trade in an input-Output framework. *Regional Studies, 41*(2), 185–196. https://doi.org/10.1080/00343400600929077.
Dietzenbacher, E. (2010). Vertical specialization in an intercountry input-output framework. *Letters in Spatial and Resource Sciences, 3*(3), 127–136. https://doi.org/10.1007/s12076-010-0043-7. Springer.
Dietzenbacher, E., et al. (2013). The construction of world input–output tables in the WIOD project. *Economic Systems Research, 25*(1), 71–98. https://doi.org/10.1080/09535314.2012.761180. Routledge.
Duchin, F. (1990). The conversion of biological materials and wastes to useful products. *Structural Change and Economic Dynamics*, *1*(2), 243–261.
Duarte, R., Pinilla, V., & Serrano, A. (2014a). The effect of globalisation on water consumption: A case study of the Spanish virtual water trade, 1849–1935. *Ecological Economics, 100,* 96–105. https://doi.org/10.1016/j.ecolecon.2014.01.020.
Duarte, R., Pinilla, V., & Serrano, A. (2014b). The water footprint of the Spanish agricultural sector: 1860–2010. *Ecological Economics, 108,* 200–207. https://doi.org/10.1016/j.ecolecon.2014.10.020.

Duarte, R., Sánchez-Chóliz, J., & Bielsa, J. (2002). Water use in the Spanish economy: An input-output approach. *Ecological Economics, 43*(1), 71–85. https://doi.org/10.1016/s0921-8009(02)00183-0.
Duarte, R., & Yang, H. (2011). Input–output and water: Introduction to the special issue. *Economic Systems Research, 23*(4), 341–351. https://doi.org/10.1080/09535314.2011.638277. Routledge.
Duchin, F. (2005). A world trade model based on comparative advantage with m regions, n goods, and k factors. *Economic Systems Research, 17*(2), 141–162. https://doi.org/10.1080/09535310500114903.
Duchin, F., & López-Morales, C. (2012). Do water-rich regions have a comparative advantage in food production?: Improving the representation of water for agriculture in economic models. *Economic Systems Research, 24*(4), 371–389.
Ercin, A. E., Aldaya, M. M., & Hoekstra, A. Y. (2011). Corporate water footprint accounting and impact assessment: The case of the water footprint of a sugar-containing carbonated beverage. *Water Resources Management, 25*(2), 721–741. https://doi.org/10.1007/s11269-010-9723-8.
Feng, K., et al. (2011). Comparison of bottom-up and top-down approaches to calculating the water footprints of nations. *Economic Systems Research, 23*(4), 371–385. https://doi.org/10.1080/09535314.2011.638276. Routledge.
Feng, K., et al. (2012). Assessing regional virtual water flows and water footprints in the Yellow River Basin, China: A consumption based approach. *Applied Geography, 32*(2), 691–701. https://doi.org/10.1016/j.apgeog.2011.08.004. Elsevier Ltd.
Feng, K., & Hubacek, K. (2015). A multi-region input-output analysis of global virtual water flows. In *Handbook of Research methods and Applications in Environmental Studies* (pp. 225–246). https://doi.org/10.4337/9781783474646.
Ferrão, P., & Nhambiu, J. (2009). A comparison between conventional LCA and Hybrid EIO-LCA: Analyzing crystal giftware contribution to global warming potential BT. In S. Suh (ed.), *Handbook of input-output economics in industrial ecology* (pp. 219–230). Dordrecht: Springer Netherlands. https://doi.org/10.1007/978-1-4020-5737-3_11.
Foster-McGregor, N., & Stehrer, R. (2013). Value added content of trade: A comprehensive approach. *Economics Letters, 120*(2), 354–357. https://doi.org/10.1016/j.econlet.2013.05.003. Elsevier B.V.
De Fraiture, C., et al. (2004). Does international cereal trade save water? *The impact of virtual water trade on global water use.*
Gawel, E., & Bernsen, K. (2011). Do we really need a water footprint? Global trade, water scarcity and the limited role of virtual water. *GAIA-Ecological Perspectives for Science and Society, 20,* 162–167.
Gawel, E., & Bernsen, K. (2013). 'What is wrong with virtual water trading? On the limitations of the virtual water concept. *Environment and Planning C: Government and Policy, 31,* 168–181.
Gerbens-Leenes, P. W. & Hoekstra, A. Y. (2008). *Business water footprint accounting: A tool to assess how production of goods and services impacts on freshwater resources worldwide, Value of Water Research Report Series*. https://doi.org/10.1016/j.ecolecon.2008.06.021.
Gerbens-Leenes, P. W., Hoekstra, A. Y., & van der Meer, T. (2009a). The water footprint of energy from biomass: A quantitative assessment and consequences of an increasing share of bio-energy in energy supply. *Ecological Economics, 68*(4), 1052–1060. https://doi.org/10.1016/j.ecolecon.2008.07.013.
Gerbens-Leenes, W., Hoekstra, A. Y., & Van Der Meer, T. H. (2009b). The water footprint of bioenergy. *Proceedings of the National Academy of Sciences of the United States of America, 106*(25), 10219–10223. https://doi.org/10.1073/pnas.0812619106.
Guan, D., & Hubacek, K. (2007). Assessment of regional trade and virtual water flows in China. *Ecological Economics, 61*(1), 159–170. https://doi.org/10.1016/j.ecolecon.2006.02.022.
Guan, D., & Hubacek, K. (2008). A new and integrated hydro-economic accounting and analytical framework for water resources: A case study for North China. *Journal of Environmental Management, 88*(4), 1300–1313. https://doi.org/10.1016/j.jenvman.2007.07.010.

Hadjikakou, M., Chenoweth, J., & Miller, G. (2013). Estimating the direct and indirect water use of tourism in the eastern Mediterranean. *Journal of Environmental Management, 114,* 548–556.

Hendricks, D. W., & Dehaan, R. W. (1981). The input-output water transactions model of supply and demand. *Water Supply and Management*, *5*(4–5), 317–330. http://www.scopus.com/inward/record.url?eid=2-s2.0-0019375176&partnerID=40.

Hoekstra, A. Y., et al. (2009a). *Water footprint manual: State of the art 2009*. Enschede, the Netherlands: Water Footprint Network.

Hoekstra, A. Y., et al. (2011). *The Water Footprint Assessment Manual: Setting the Global Standard*. London, UK.

Hoekstra, A. Y. (2013). The water footprint of modern consumer society. *Water Resources Management, 27*(11), 3847–3848. https://doi.org/10.1007/s11269-013-0409-x.

Hoekstra, A. Y., et al. (no date) *Water footprint manual: State of the art (2009, 2011), Water Footprint Network*. Enschede, the Netherlands.

Hoekstra, A. Y., & Chapagain, A. K. (2007). Water footprints of nations: Water use by people as a function of their consumption pattern. *Water Resources Management, 21*(1), 35–48. https://doi.org/10.1007/s11269-006-9039-x.

Hoekstra, A. Y., Gerbens-Leenes, W., & Van Der Meer, T. H. (2009) Reply to Pfister and Hellweg: Water footprint accounting, impact assessment, and life-cycle assessment. *Proceedings of the National Academy of Sciences of the United States of America, 106*(40). https://doi.org/10.1073/pnas.0909948106.

Hoekstra, A. Y., & Hung, P. Q. (2002) Virtual water trade: A quantification of virtual water flows between nations in relation to international crop trade. *Virtual Water Trade: A Quantification of Virtual Water Flows Between Nations in Relation to International Crop Trade.*

Hoekstra, A. Y., & Mekonnen, M. M. (2012). The water footprint of humanity. *Proceedings of the National Academy of Sciences, 109*(9), 3232–3237. https://doi.org/10.1073/pnas.1109936109.

Hubacek, K., et al. (2016). Linking local consumption to global impacts. *Journal of Industrial Ecology*, *0*(0), 382–386. https://doi.org/10.1111/jiec.12463.

Hummels, D., Ishii, J., & Yi, K.-M. (1999). *The nature and growth of vertical specialization in world trade.*

Isard, W. (1951). Interregional and regional input-output analysis: A model of a space economy. *Review of Economics and Statistics, 33*(4), 318–328.

Kondo, Y., & Nakamura, S. (2003). Decision analytic extension of waste input–output model based on linear programming. In *Paper presented at the second meeting of the international society for industrial ecology*. Michigan: Ann Arbor.

Koopman, R., Wang, Z., & Wei, S.-J. (2014). Tracing value-added and double counting. *American Economic Review, 104*(2), 459–494. https://doi.org/10.1257/aer.104.2.459.

Krzywinski, M. I., et al. (2009). Circos: An information aesthetic for comparative genomics. *Genome Research*. https://doi.org/10.1101/gr.092759.109.

Lenzen, M. (2009). Understanding virtual water flows: A multiregion input-output case study of Victoria. *Water Resources Research, 45*. https://doi.org/10.1029/2008wr007649.

Lenzen, M. (2011). Aggregation versus disaggregation in input–output analysis of the environment. *Economic Systems Research, 23*(1), 73–89. https://doi.org/10.1080/09535314.2010.548793. Routledge.

Lenzen, M., et al. (2013a). International trade of scarce water. *Ecological Economics, 94,* 78–85. https://doi.org/10.1016/j.ecolecon.2013.06.018.

Lenzen, M., Moran, D., & Kanemoto, K. (2013b). Building EORA: A global multi- region input–output database at high country and sector resolution. *Economic Systems Research, 25*(1), 37–41.

Leontief, W. (1970). Environmental repercussions and the economic structure: an input-output approach. *The Review of Economics and Statistics*, *52*(3), 262–271.

Lofting, E. M., & McGauhey, P. H. (1968). *Economic valuation of water. An input-output analysis of California water requirements.*

Lopez-Gunn, E., & Llamas, M. R. (2008). Re-thinking water scarcity: Can science and technology solve the global water crisis? *Natural Resources Forum, 32*(3), 228–238. https://doi.org/10.1111/j.1477-8947.2008.00200.x.
López-Morales, C. A., & Duchin, F. (2014). Economic implications of policy restrictions on water withdrawals from surface and underground sources. *Economic Systems Research*, 1–18. https://doi.org/10.1080/09535314.2014.980224. Routledge.
López-Morales, C., & Duchin, F. (2011). Policies and technologies for a sustainable use of water in Mexico: a scenario analysis. *Economic Systems Research, 23*(4), 387–407. https://doi.org/10.1080/09535314.2011.635138. Routledge.
Makate, C., Wang, R., & Tatsvarei, S. (2018). Water footprint concept and methodology for warranting sustainability in human-induced water use and governance. *Sustainable Water Resources Management, 4*(1), 91–103. https://doi.org/10.1007/s40899-017-0143-2.
Mason, N. (2013). *Uncertain frontiers: mapping new corporate engagement in water security*. London.
Mekonnen, M. M., & Hoekstra, A. Y. (2010). The green, blue and grey water footprint of farm animals and animal products, 1(48). https://doi.org/10.5194/hess-15-1577-2011.
Mekonnen, M. M., & Hoekstra, A. Y. (2011a). *Green, blue and grey water footprint of production and consumption*. Delft, The Netherlands: UNESCO-IHE. https://doi.org/10.5194/hessd-8-763-2011.
Mekonnen, M. M., & Hoekstra, A. Y. (2011b). *The blue water footprint of electricity from hydropower*.
Mekonnen, M. M., & Hoekstra, A. Y. (2011c). The green, blue and grey water footprint of crops and derived crop products. *Hydrology and Earth System Sciences, 15*(5), 1577–1600. https://doi.org/10.5194/hess-15-1577-2011. Copernicus Publications.
Milà i Canals, L., et al. (2009) Assessing freshwater use impacts in LCA: Part I-Inventory modelling and characterisation factors for the main impact pathways. *International Journal of Life Cycle Assessment, 14*(1), 28–42. https://doi.org/10.1007/s11367-008-0030-z.
Miller, R. E., & Blair, P. D. (2009). *Input-output analysis: Foundations and extensions, Cambridge Books from Cambridge University*. Cambridge, UK: Cambridge University Press. http://books.google.com/books?id=PVoZPQAACAAJ.
Moran, D., & Wood, R. (2014). Convergence between the EORA, WIOD, EXIOBASE, and OPENEU's consumption-based carbon accounts. *Economic Systems Research. Routledge, 26*(3), 245–261. https://doi.org/10.1080/09535314.2014.935298.
Morishima, M. (1973). *Marx's economics*. Cambridge University Press.
Moses, L. N. (1955). The stability of interregional trading patterns and input-output analysis. *American Economic Review, 45*, 803–832.
Nakamura, S. (1999). Input–output analysis of waste cycles. In *Proceedings of the first international symposium on environmentally conscious design and inverse manufacturing* (pp. 475–480). Los Alamitos: IEEE Computer Society.
Nakamura, S., & Kondo, Y. (2002). Input-output analysis of waste management. *Journal of Industrial Ecology, 6*(1), 39–63.
Narayanan, G., Aguiar, A., & McDougall, R. (2012). *Global trade, assistance, and production: The GTAP 8 data base*. Center for Global Trade Analysis, Purdue University.
Narayanan, G., Badri, A. A., & McDougall, R. (2015). *Global trade, assistance, and production: The GTAP 9 data base*. Center for Global Trade Analysis, Purdue University.
OECD (2016). OECD Inter-Country Input-Output (ICIO). Organization for Economic Cooperation and Development.
Okadera, T., Watanabe, M., & Xu, K. (2006). Analysis of water demand and water pollutant discharge using a regional input–output table: An application to the City of Chongqing, upstream of the Three Gorges Dam in China. *Ecological Economics, 58*(2), 221–237. https://doi.org/10.1016/j.ecolecon.2005.07.005.

Oki, T., et al. (2003). Virtual water trade to Japan and in the world. In A. Y. Hoekstra (Ed.), *Virtual water trade: proceedings of the international expert meeting on virtual water trade. Value of water research report series n 12*. Delft, The Netherlands: UNESCO-IHE.

Pellicer-Martínez, F., & Martínez-Paz, J. M. (2016). The water footprint as an indicator of environmental sustainability in water use at the river basin level. *Science of the Total Environment*, *571*, 561–574. https://doi.org/10.1016/j.scitotenv.2016.07.022. Elsevier B.V.

Perry, C. (2014). Water footprints: Path to enlightenment, or false trail? *Agricultural Water Management, 134,* 119–125. https://doi.org/10.1016/j.agwat.2013.12.004.

Pfister, S., et al. (2017). Understanding the LCA and ISO water footprint: A response to Hoekstra (2016) "A critique on the water-scarcity weighted water footprint in LCA". *Ecological Indicators*, *72*(Supplement C), 352–359. https://doi.org/10.1016/j.ecolind.2016.07.051.

Pfister, S., Koehler, A., & Hellweg, S. (2009). Assessing the environmental impacts of freshwater consumption in LCA. *Environmental Science & Technology*, *43*(11), 4098–4104. http://www.ncbi.nlm.nih.gov/pubmed/19569336.

Ridoutt, B. G., et al. (2009). Water footprinting at the product brand level: case study and future challenges. *Journal of Cleaner Production, 17*(13), 1228–1235. https://doi.org/10.1016/j.jclepro.2009.03.002.

Ridoutt, B. G., & Pfister, S. (2010). A revised approach to water footprinting to make transparent the impacts of consumption and production on global freshwater scarcity. *Global Environmental Change, 20*(1), 113–120. https://doi.org/10.1016/j.gloenvcha.2009.08.003.

Rivas Ibáñez, G., et al. (2017). A corporate water footprint case study: The production of Gazpacho, a chilled vegetable soup. *Water Resources and Industry, 17,* 34–42. https://doi.org/10.1016/j.wri.2017.04.001.

Roemer, J. (1982) *A general theory of exploitation and class*. Harvard University Press.

Sánchez-Chóliz, J., & Duarte, R. (2000). The economic impacts of newly irrigated areas in the Ebro Valley. *Economic Systems Research. Routledge, 12*(1), 83–98. https://doi.org/10.1080/095353100111290.

Sánchez-Chóliz, J., & Duarte, R. (2005). Water pollution in the Spanish economy: Analysis of sensitivity to production and environmental constraints. *Ecological Economics, 53*(3), 325–338. https://doi.org/10.1016/j.ecolecon.2004.09.013.

Sánchez Chóliz, J., Arrojo, P., & Bielsa, J. (1995). Water values for Aragon. In L. A. Albisu & C. Romero (Eds.), *Environmental and Land Issues* (pp. 475–489). Wissenschaftsuerlag Vauk Kiel: Kiel.

Schendel, E. K., et al. (2007). Virtual water: A framework for comparative regional resource assessement. *Journal of Environmental Assessment Policy and Management*, *9*(3), 341–355. http://www.scopus.com/inward/record.url?eid=2-s2.0-35649005417&partnerID=40.

Serrano, A., et al. (2016). Virtual water flows in the EU27: A consumption-based approach. *Journal of Industrial Ecology, forthcomin*(3), 547–558. https://doi.org/10.1111/jiec.12454.

Shi, C., & Zhan, J. (2015). An input-output table based analysis on the virtual water by sectors with the five northwest provinces in China. *Physics and Chemistry of the Earth, 79–82,* 47–53. https://doi.org/10.1016/j.pce.2015.03.004. Elsevier Ltd.

Stehrer, R. (2012). Value added trade: A tale of two concepts.

Suh, S. (2004a). A note on the calculus for physical input-output analysis and its application to land appropriation of international trade activities. *Ecological Economics*, *48*(1), 9–17. http://www.scopus.com/inward/record.url?eid=2-s2.0-0742320048&partnerID=40.

Suh, S. (2004b). Functions, commodities and environmental impacts in an ecological-economic model. *Ecological Economics, 48*(4), 451–467. https://doi.org/10.1016/j.ecolecon.2003.10.013.

Suh, S., et al. (2004). System boundary selection in life-cycle inventories using hybrid approaches. *Environmental Science and Technology, 38*(3), 657–664. https://doi.org/10.1021/es0263745.

Suh, S., & Huppes, G. (2005). Methods for life cycle inventory of a product. *Journal of Cleaner Production, 13*(7), 687–697. https://doi.org/10.1016/j.jclepro.2003.04.001.

Suh, S., & Kagawa, S. (2005). Industrial ecology and input-output economics: An introduction. *Economic Systems Research, 17*(4), 349–364. https://doi.org/10.1080/09535310500283476.

Suh, S. W., & Heijungs, R. (2007). Power series expansion and structural analysis for life cycle assessment. *International Journal of Life Cycle Assessment, 12*(6), 381–390. https://doi.org/10.1065/lca2007.08.360.
Tukker, A., et al. (2009). Towards a global multiregional environmentally-extended input-output database. *Ecological Economics, 68*(7), 1928–1937.
Tukker, A., et al. (2013). EXIOPOL–Development and illustrative analyses of a detailed global MR EE SUT/IOT. *Economic Systems Research. Routledge, 25*(1), 50–70. https://doi.org/10.1080/09535314.2012.761952.
Tukker, A., & Dietzenbacher, E. (2013). Global multiregional input–output frameworks: an introduction and outlook. *Economic Systems Research. Routledge, 25*(1), 1–19. https://doi.org/10.1080/09535314.2012.761179.
Vanham, D., & Bidoglio, G. (2013). A review on the indicator water footprint for the EU28. *Ecological Indicators, 26,* 61–75. https://doi.org/10.1016/j.ecolind.2012.10.021.
Vanham, D., Mekonnen, M. M., & Hoekstra, A. Y. (2013). The water footprint of the EU for different diets. *Ecological Indicators, 32,* 1–8. https://doi.org/10.1016/j.ecolind.2013.02.020.
Vegara, J. (1979). *Economía política y modelos multisectoriales*. Madrid: Tecnos.
Velázquez, E. (2006). An input-output model of water consumption: Analysing intersectoral water relationships in Andalusia. *Ecological Economics, 56*(2), 226–240. https://doi.org/10.1016/j.ecolecon.2004.09.026.
Vélazquez, E. (2001). *Consumo de Agua y la Contaminación Hídrica en Andalucía. Un análisis desde el Modelo Input-Output y la Teoría de Grafos*. PhD Thesis. Universidad Pablo Olavide de Sevilla (España).
Vörösmarty, C. J., et al. (2015). Fresh water goes global. *Science, 349*(6247), 478 LP-479. http://science.sciencemag.org/content/349/6247/478.2.abstract.
Wang, X., et al. (2016). An input-output structural decomposition analysis of changes in sectoral water footprint in China. *Ecological Indicators, 69,* 26–34. https://doi.org/10.1016/j.ecolind.2016.03.029. Elsevier.
Wang, Y., Xiao, H. L., & Lu, M. F. (2009). Analysis of water consumption using a regional input-output model: Model development and application to Zhangye City, Northwestern China. *Journal of Arid Environments, 73*(10), 894–900. https://doi.org/10.1016/j.jaridenv.2009.04.005.
Wichelns, D. (2010). Virtual water: A helpful perspective, but not a sufficient policy criterion. *Water Resources Management, 24*(10), 2203–2219. https://doi.org/10.1007/s11269-009-9547-6.
Wichelns, D. (2011). Virtual water and water footprints: Compelling notions, but notably flawed: reaction to two articles regard the virtual water concept. *GAIA, 20,* 171–175.
Wichelns, D. (2015). Virtual water and water footprints: Overreaching into the discourse on sustainability, efficiency, and equity. *Water Alternatives, 8*(3), 396–414.
Wood, R., et al. (2015). Global sustainability accounting—Developing EXIOBASE for multi-regional footprint analysis. *Sustainability*. https://doi.org/10.3390/su7010138.
Yang, H., Pfister, S., & Bhaduri, A. (2013). Accounting for a scarce resource: virtual water and water footprint in the global water system. *Current Opinion in Environmental Sustainability, 5*(6), 599–606. https://doi.org/10.1016/j.cosust.2013.10.003.
Younger, P. L. (2006). The water footprint of mining operations in space and time-A new paradigm for sustainability assessments?, in *Australasian Institute of Mining and Metallurgy Publication Series*, pp. 13–21. http://www.scopus.com/inward/record.url?eid=2-s2.0-58049164296&partnerID=40.
Yu, Y., et al. (2010). Assessing regional and global water footprints for the UK. *Ecological Economics, 69*(5), pp. 1140–1147. https://doi.org/10.1016/j.ecolecon.2009.12.008. Elsevier B.V.
Zhang, C., & Anadon, L. D. (2014). A multi-regional input-output analysis of domestic virtual water trade and provincial water footprint in China. *Ecological Economics, 100*, 159–172. https://doi.org/10.1016/j.ecolecon.2014.02.006. Elsevier B.V.
Zhang, Y., et al. (2017). Mapping of water footprint research: A bibliometric analysis during 2006–2015. *Journal of Cleaner Production, 149*(Supplement C), 70–79. https://doi.org/10.1016/j.jclepro.2017.02.067.

Zhao, X., et al. (2010). Applying the input-output method to account for water footprint and virtual water trade in the Haihe River basin in China. *Environmental Science and Technology, 44*(23), 9150–9156. https://doi.org/10.1021/es100886r.

Zhao, X. (2014). China's inter-regional trade of virtual water: A multi-regional Input-output modeling, (pp. 1–27).

Zhao, X., Chen, B., & Yang, Z. F. (2009). National water footprint in an input-output framework-A case study of China 2002. *Ecological Modelling, 220*(2), 245–253. https://doi.org/10.1016/j.ecolmodel.2008.09.016.

Zimmer, D., & Renault, D. (2003). Virtual water in food production and global trade: review of methodological issues and preliminary results. In A. Y. Hoekstra (Ed.) *Virtual Water Trade: Proceedings of the International Expert Meeting on Virtual Water Trade*. Delft, The Netherlands: Value of Water Research Report Series n 12. UNESCO-IHE.

Water Footprint of a Decentralised Wastewater Treatment Strategy Based on Membrane Technology

A. Arias, I. Vallina, Y. Lorenzo, O. T. Komesli, E. Katsou, G. Feijoo and M. T. Moreira

Abstract Growing pressure on water resources has led to the search for alternatives to conventional wastewater treatment plants (WWTPs). Centralized wastewater treatment systems provide a single treatment scheme but are not especially adequate for water reuse as large flows of reclaimed water need to be efficiently managed. As an alternative, a new concept of wastewater treatment based on decentralized systems arise, which comprises collection, treatment and final disposal and/or reuse of water in an area close to the point of origin. Turkey is severely affected by water scarcity, thus, it is essential to improve water recovery through efficient technologies that allow nutrient recovery and have the potential for water reuse for irrigation to counteract consumption of drinking water. In this study, a decentralized membrane bioreactor (MRB) plant in Turkey was evaluated within the framework of the most relevant environmental indicators under the approach of Life Cycle Assessment: climate change (CC) and eutrophication potential (EP). In addition, the water scarcity footprint indicator was estimated using the available remaining water method (AWARE). This category should be taken into account when addressing the potential benefits associated with water reuse. Once the impacts of the plant under study were determined, a sensitivity analysis was carried out considering different solid retention times (SRT) in the MBR operation and the influence of the impacts associated with the construction phase. The sub-processes with the greatest impacts are electricity consumption in the operational phase and infrastructure in the construction phase. These impacts are significantly reduced when water is reused for irrigation of green areas, approximately 23% in CC, 4.8% in EP and 133.8% in AWARE indicator. No significant influence of the SRT variable was observed on environmental impacts for

A. Arias · I. Vallina · Y. Lorenzo · G. Feijoo · M. T. Moreira (✉)
Department of Chemical Engineering, University of Santiago de Compostela, 15782 Santiago de Compostela, Spain
e-mail: maite.moreira@usc.es

O. T. Komesli
Department of Environmental Engineering, Ataturk University, Erzurum, Turkey

E. Katsou
Department of Civil and Environmental Engineering, Brunel University London, Uxbridge, UK

S. S. Muthu (ed.), *Environmental Water Footprints*, Environmental Footprints and Eco-design of Products and Processes,
https://doi.org/10.1007/978-981-13-2508-3_4

the range examined, since it only affected the eutrophication category, determining an optimum SRT value of 50 days for the MBR operation.

Keywords Decentralized scale · Life cycle methodology (LCA) Membrane bioreactor (MBR) · Sensitivity analysis · Water scarcity footprint Water reuse

Nomenclature

AD	Anaerobic Digestion
AMD	Availability Minus Demand
AnMBR	Anaerobic Membrane Bioreactor
ASP	Activated Sludge Process
AWARE	Available Water Remaining
BW	Black Water
BWS	Black Water Source-Separation
CAS	Conventional Activated Sludge
CC	Climate Change
CF	Characterization Factor
CML	Centre of Environmental Science of Leiden University
COD	Chemical Oxygen Demand
CP	Composting
CTA	Consumption to Availability
DTA	Demand to Availability
EB	Environmental Burdens
EP	Eutrophication Potential
EPDM	Synthetic rubber
EWR	Environmental Water Requirements
FAT	Full-advanced treatment
FU	Functional Unit
GAC	Granular Activated Carbon
GHG	Greenhouse gases
GROW	Green Roof Water
GW	Grey Water
HDPE	High Density Polyethylene
HRAS	High Rate Activated Sludge
HRT	Hydraulic Retention Time
HS	Hybrid System
HWC	Human Water Consumption
IPCC	Intergovernmental Panel on Climate Change
LDPE	Low Density Polyethylene
LCA	Life Cycle Assessment
LCI	Life Cycle Inventory

LF	Landfilling
MBR	Membrane Bioreactor
MCF	Membrane Chemical Reactor
MF	Microfiltration membrane
MFZ	Multi-family zone
MLSS	Mixed Liquor Suspended Solids
NEB	Net Environmental Benefit
NR	Natural Resources
PEI	Potential Environmental Impacts
PES	Polyethersulfone
PET	Polyethylene Terephthalate
PVDF	Polyvinylidene fluoride
SBR	Sequencing Batch Reactor
SFZ	Single-family zone
SGD	Specific Gas Demand
SRT	Solids Retention Time
TMP	Transmembrane pressure
UASB	Upflow Anaerobic Sludge Blanket
UFO	Ultrafiltration osmotic
UK	United Kingdom
UV	Ultraviolet
VFM	Variable Flow Method
VRM	Membrane Vacuum Bioreactor
WFN	Water Footprint Network
WTA	Withdrawal to Availability
WULCA	Water Use in LCA
WWTP	Wastewater treatment plant

1 Introduction

Although more than 70% of the surface of Earth is covered by water, only 1% of the Earth's water is available as fresh water. Most of this water is used in agriculture and industry and only 10% is drinking water. Reduced water availability is related to economic, social, political and cultural aspects such as population growth and climate change (Kounina et al. 2013). In recent years, water and energy consumption have increased significantly. On the other hand, the drop in the rainfall collected in combination with high evaporation contributes to water scarcity. Therefore, integrated water resources management is one of the main aspects from the perspective of the urban, agricultural and industrial water cycle. In this context, water reuse has emerged as the most viable alternative as reclaimed water reduces the demand for fresh water (Hochstrat et al. 2007). The reuse scheme in WWTPs is considered an increasingly important factor in the Water Framework Directive, which aims to make WWTPs more efficient and less costly (Bixio et al. 2006).

The centralized wastewater management strategy was developed in the mid-19th century (Hophmayer-Tokich 2006). It is composed of a sewerage system that collects all wastewater from households, commercial areas and industries, its treatment in a centralized facility and the discharge of the treated effluent far from the point of origin, making it difficult to reuse as process or irrigation water (Chen and Wang 2009; Hophmayer-Tokich 2006). Conventional systems present high investment and maintenance costs as they require the construction of extensive collection and treatment infrastructures (Remy and Jekel 2008). For this reason, in areas characterized by water scarcity or socio-economic instability, the centralized approach is not considered the most viable option (Meuler et al. 2008). In return, decentralized schemes emerge as alternatives of growing interest that include the final disposal and/or reuse of reclaimed water in an area close to the point of origin (Hophmayer-Tokich 2006; Massoud et al. 2009; Meuler et al. 2008). Therefore, decentralized systems that consider closed circuits of water and nutrients for the reuse of these resources can play an important role in sustainable water management. It is clear that the decentralized system must ensure a lower cost associated with infrastructure and electricity consumption. This is beneficial for communities, primarily in developing countries, which have technological and economic constraints and limited availability of resources (Massoud et al. 2009). As these systems are more compact, it allows the installation of more advanced technologies, such as membrane bioreactors, which adapt the quality of reclaimed water to local needs (Prieto et al. 2013).

In this context, Mediterranean countries have experienced increasing water scarcity in recent decades. In particular, annual per capita water availability in Turkey is about one-fifth of that in water-rich countries, which is below the world average (Yuksel 2015). Therefore, Turkey is urged to improve water availability considering the foreseen estimations for 2023, when the amount of water available will be less than 1,000 m^3/(inhab·year) (Arslan-Alaton et al. 2010; Yuksel 2015). In Table 1, several countries were classified according to their water availability. In addition, water demand in Turkey is steadily increasing in the agricultural sector, with a demand of about 34 billion m^3 per year in 2007, representing 74% of total consumption (Cakmak et al. 2007). Figure 1 shows the water consumption in Turkey in 2012 and the forecasted values for 2023. With a forecasted water availability of about 72 mil millions m^3/year in 2013 (Yuksel 2015), water reserves will not meet the demand.

The management of water resources arises as a major problem. In Turkey, water reuse for irrigation has been carried out through direct use (untreated or partially treated wastewater) or after being mixed with river water. Accordingly, this water is not suitable for agricultural use since it does not comply with the quality criteria for water reuse such as total coliforms, conductivity and salinity (Arslan-Alaton et al. 2010). Thus, with the aim of changing the existing strategy for the reuse of partially treated wastewater, a paradigm shift is needed to consider technologies that ensure high quality of the treated effluent (Arslan-Alaton et al. 2010). Conventional wastewater treatment is not entirely satisfactory, as many compounds have been found to persist unchanged after tertiary treatment. Therefore, it is necessary to develop advanced treatment technologies oriented to the elimination of specific contaminants, in addition to microorganisms and colloids (Gil et al. 2012). These

Table 1 Classification by water availability per capita of several countries of the world

Classification	Country	Area (km^2)[a]	Population (inh.)[a]	Water resources[a] (km^3/year)	Water availability (m^3/inh·year)
1	Brazil	8,515,700	207,847,528	8647.0	41,603
2	Georgia	69,700	3,999,812	63.3	15,833
3	Vietnam	330,970	93,447,601	884.1	9.461
4	EE. UU	9,831,510	321,773,631	3069.0	9.538
5	Thailand	513,120	97,959,359	438.6	6,454
6	Japan	377,960	126,573,481	430.0	3,397
7	France	549,090	64,395,345	211.0	3,277
8	Turkey	785,350	78,665,830	211.6	2,690
9	Spain	505,940	46,121,699	111.5	2,418
10	UK	243,610	64,395,810	147.0	2,271
11	China	9,600,010	9,600,010	2840.0	2,018
12	Germany	357,380	80,688,545	154.0	1,909
13	Iran	1,745,150	79,109,272	137.0	1,732
14	South Africa	1,219,090	54,490,406	51.4	942
15	Egypt	1,001,450	91,508,084	58.3	637

[a](FAO 2018)

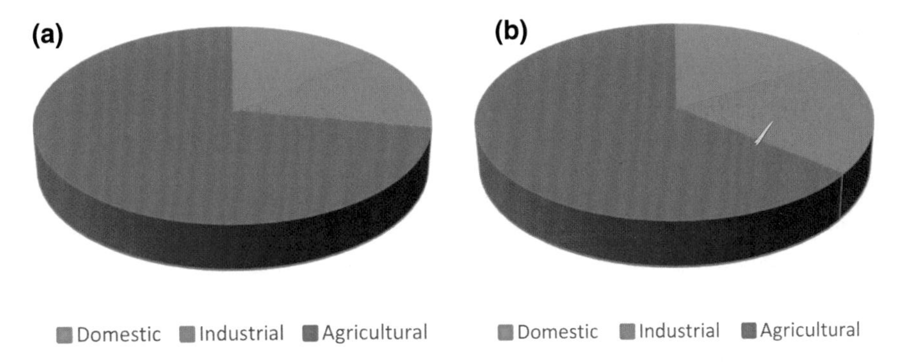

Fig. 1 **a** Water consumption in Turkey according to the use in 2012 **b** water availability in Turkey in 2023. Adapted from: (DSI 2012)

advanced technologies are based on physico-chemical, biological o hybrid processes. Additionally, physico-chemical technologies can be classified into: (i) adsorption (ii) membrane technologies and (iii) advanced oxidation processes. These technologies are explained below:

Adsorption is a superficial phenomenon that consists of increasing the concentration of a certain compound on the surface of a solid. This technology has great potential for the treatment of secondary effluent. It is also effective in the elimination of micropollutants (Luo et al. 2014).

Membrane technologies are characterized by high selectivity and efficiency to obtain a high quality effluent that meets the most demanding reuse requirements. There are several types of membrane technologies: (i) microfiltration, based on the use of symmetrical porous membranes with a pore diameter between 0.1 and 1.0 μm to remove turbidity and suspended particles. This technique is mainly applied in pharmaceutical industries (Wang et al. 2011). (ii) Ultrafiltration uses porous membrane with an average size of 0.001–0.1 μm that allows the separation of macromolecules, colloids and low molecular weight compounds (Chen et al. 2006) (iii) Nanofiltration operates at the interface of porous and non-porous membranes with average pore diameter of 0.001 μm which allows high permeability for monovalent salts (NaCl or KCl). The main disadvantage is the exposure to scaling and incrustations that may imply a decrease in permeate flow (Chen et al. 2006). Finally, the fourth option considers reverse osmosis, where it is possible to separate a solvent from a concentrated solution using a semi-permeable membrane operating at a pressure higher than the osmotic pressure. It is a highly selective process that provides removal of a wide variety of emerging contaminants, including metals and organic micro-pollutants. The main disadvantage is the high operational cost associated with energy consumption (Coday et al. 2014).

Advanced oxidation processes are based on the interaction of highly reactive species such as hydroxyl radicals. These processes include photocatalysis based on ultraviolet radiation, ozonation, Fenton process, ultrasounds, among others (Comninellis et al. 2008).

However, wastewater reclamation can be complex, costly and resource-intensive due to the need of implementing advanced treatment processes to achieve the required effluent quality. The selection of one or another process for treating wastewater can be economically and environmentally complex (Pintilie et al. 2016). In addition, certain environmental indicators such as GHG emissions, energy demand or carbon footprint could present higher values compared to only discharging the untreated effluent into the environment. In this context, the biological treatments as the membrane bioreactor (MBR) technology has been widely applied for wastewater treatment due to the advantages it offers, such as its compact and simple design, adequate biomass control, high hydraulic efficiency, the possibility of increasing treatment capacity, as well as the high quality of the effluent (Brepols et al. 2008; Komesli et al. 2015). In addition, the MBR system is particularly interesting for decentralized systems as it allows the reuse of the treated effluent for irrigation of agricultural land or recharge of aquifers (Brepols et al. 2008). There are three types of membrane: hollow fiber, flat and tubular, with different types of materials such as PES, PVDF and others. In the development of membrane technology, submerged MBR is the option that presents the least energy consumption. In terms of materials, despite the predominance of hollow fiber, flat sheet has many advantages, high flux rate, no need for backwash and reduced chemical cleaning stages. The selection of the most appropriate membrane configuration should depend on the scale of application, technical feasibility and environmental impact. The main disadvantage of MBR is

Table 2 Wastewater treatment plants (WWTPs) with MBR technology in Europe

Plant	Implementation	Location	Capacity (hab-eq)	Membrane type
Schilde[a, b, c]	2003	Belgium	10,000	Hollow fibre
Eitorf[b]	2005	Germany	7,500	Flat sheet
Bergheim-Glessen[a, b]	2007	Belgium	9,000	Hollow fibre
Brescia[b]	2002	Italy	46,000	Hollow fibre
Viareggio[d]	n.d	Italy	100,000	Hollow fibre
Heenuliet[b]	2006	Netherlands	3,300	Flat sheet
Rietliau[a]	2005	Switzerland	44,000	n.d

[a]Brepols et al. (2008)
[b]Judd (2011)
[c]Wiessmann et al. (2007)
[d]Battistoni et al. (2006)

that the energy consumption (0.64 kWh/m^3 permeate) is higher than conventional systems (0.3 kWh/m^3) due mainly to membrane fouling. Mechanical and chemical cleaning is therefore carried out (Brepols et al. 2008; Bui et al. 2016; Krzeminski et al. 2012). In the last 15 years, membrane bioreactors have begun to be implemented in many wastewater treatment plants. Some of the industrially installed are shown in Table 2.

One of the most outstanding tools in the topic of environmental assessment is the Life Cycle Assessment (LCA) methodology, described in ISO 14040 standards. There are different references that analyze the environmental profile of the decentralized approach. Nogueira et al. (2009) analyzed two decentralized systems with low energy consumption: (i) constructed wetland and (ii) slow infiltration and a conventional system with an activated sludge process from environmental and economic perspectives taking into account the operational and construction phase. They concluded that decentralized systems have enormous potential associated with energy savings and environmental improvements, so these technologies should not be discarded. Shehabi et al. (2012) compared a conventional system and a decentralized system in California. In both systems, energy consumption, greenhouse gas (GHG) emissions and effluent quality were compared. They concluded that conventional systems help to reduce the pollution burden by at least one-fifth less compared to decentralized systems for the same volume treated. The results show that decentralized systems must be carefully evaluated as part of the design process. Lehtoranta et al. (2014) analyzed six decentralized wastewater scenarios in rural communities in Finland. The scenarios studied were: (i) sequential batch reactor (SBR), (ii) biofilters, (iii) soil infiltration, (iv) sand filter, (v) black water tank (BW) and soil infiltration for grey water (GW) and (vi) dry toilets with GW treatment. They concluded that dry toilets with GW treatment had the lowest environmental impact. However, the optimal solution depends on local conditions and appropriate guidance is needed.

Thibodeau et al. (2014) compared the environmental performance of a conventional system and a black water source separation system (BWS). BWS had more impacts on the categories of human health, climate change and ecosystem quality. Therefore, the conclusion of this study was that BWS systems need significant improvements to obtain better environmental conditions than conventional systems. Matos et al. (2014) compared a decentralized system that considers the reuse of the treated effluent with a centralized system. The system was designed for a population equivalent of 5,000 inhabitants and evaluated in terms of water quality, CO_2 emissions and energy consumption. The construction phase was not included in the system boundaries. The study determined that for the centralized system, CO_2 emissions were between 329–4879 kg CO_2/d, while electricity consumption is around 886–13215 kWh/d. In the decentralized system, CO_2 emissions were between 123 and 7575 kg CO_2/d and electricity consumption varied between 335 and 1559 kWh/day. Thus, Matos et al. (2014) determined that the decentralized system presented better environmental profile than the centralized system. Lam et al. (2015) conducted an assessment of four different scenarios: (i) offsite treatment, (ii) onsite treatment, (iii) source-separation and (iv) pour-flush toilet use. The study was conducted in a hypothetical location in the municipality of Tianjin, China. The conclusion of this study was that the system with separation (scenario iii) presented the best results in terms of direct water use in all impact categories. In another similar work, Opher and Friedler (2016) conducted a study to compare the impacts of four alternatives for a hypothetical water supply and wastewater treatment scheme. The first two alternatives were based on a conventional system with the difference that the second alternative included the reuse of treated wastewater for the irrigation of green areas. The other two were based on decentralized systems for the treatment of black and grey water, with the latter being used to irrigate green areas. The conclusion was that decentralized systems had fewer impacts on the impact categories, since the main source of impact was the energy consumption of conventional systems. Jeong et al. (2018) also evaluated a hybrid system (HS) between conventional and decentralized systems in Atlanta (USA). The authors performed a simulation in which five single-family houses (SFZ) and four dwellings of multi-family apartment buildings (MFZ) were evaluated. They concluded that the HS had less environmental impact and that the demand for energy and drinking water could be reduced. Moreover, the results showed that in the SFZ, water reuse would be higher than in the MFZ, as the latter considered less irrigation of green areas. Average reductions of 20% in electricity consumption could also be achieved (17–49% in SFZ, 32–41% in MFZ). In addition, the LCA score for the recovery system was 20–41% lower than that of the current centralized system, mainly due to lower electricity consumption, with scores stabilizing at higher population densities.

Finally, Dominguez et al. (2018) assessed the treatment of the laundry water of a 70 guest hotel in Santander (Cantabria, Spain). They compared three reuse scenarios: (i) photocatalysis (ii) photovoltaic solar-driven photocatalysis and (iii) MBR. Natural resources (NR) and environmental burdens (EB) were the environmental indicators considered to report the results. The MBR had the greatest impact on NR due to energy consumption, but lower emissions to air. Regarding photovoltaic solar energy-driven photocatalysis, this option had less environmental impacts and should be considered as an alternative in the decentralized systems.

Focusing on LCA reports for MBR systems, Tangsubkul et al. (2005) evaluated the environmental profile of three different wastewater treatment schemes: (i) ozonation followed by microfiltration and chemical disinfection, (ii) membrane bioreactor followed by reverse osmosis and (iii) stabilization lagoon. Despite the low indirect environmental impacts of the latter option, the treatment efficiency was higher for MBR systems compared to other alternatives. Tangsubkul et al. (2006) analyzed the effect that operational conditions had on the environmental profile of a microfiltration (MF) membrane. The best environmental profile occurred when the MF was operated at low flux and high maximum trans-membrane pressure. In addition, the sensitivity analysis indicated that within the low flux range, the chemical cleaning frequency could affect the environmental performance of the entire MF process. Ortiz et al. (2007) compared a conventional activated sludge (CAS) system, alone or coupled with an ultrafiltration membrane as tertiary treatment, a submerged MBR system and an external MBR. The submerged MBR had lower environmental impacts than the conventional system followed by ultrafiltration and the external MBR system. The authors concluded that the implementation of membranes provided higher effluent quality but also high environmental impact, mainly due to the indirect emissions associated to the energy consumption. Memon et al. (2007) analyzed four wastewater treatment technologies: (i) MBR (ii) membrane chemical reactors (MCR) (iii) reed beds and (iv) an innovative green roof water recycling system (GROW). They concluded that the construction phase was not very relevant for the MBR, but that the environmental impacts related to the operational phase were high due to the power consumption of these reactors. Wenzel et al. (2008) evaluated different wastewater treatments: (i) sand filtration (ii) ozone treatment and (iii) MBRs with the objective of studying the removal of the pathogens and micro-pollutants. They concluded that the sand filtration achieved better removal of heavy metals. Moreover, sand filtration showed a net environmental benefit for the assumptions in this study. Høibe et al. (2008) analyzed five advanced technologies: (i) sand filtration (ii) ozone (iii) ultraviolet (UV) disinfection (iv) MBR (v) UV+ advanced oxidation. They concluded that the best technology from an environmental and technological perspective is sand filtration due to low energy consumption and removal of heavy metals. Fenu et al. (2010) analyzed the energy requirements related to the cost for the full-scale MBR compared to the conventional activated systems. The conclusion to this study was that the cost of MBR process (0.64 kWh/m^3 of permeate) was higher than for CAS reactors (0.3 kWh/m^3). In a previous study, the operation of four different biological reactors, all of them coupled with a hollow fiber unit, were assessed: (i) an activated sludge reactor; (ii) anoxic and aerobic tanks; (iii) a sequence of anaerobic reactor, two anoxic tanks and one final aerobic reactor and (iv) UASB rector followed by a hybrid reactor. The main contributors to the environmental impacts were identical for the four configurations: electricity consumption and sludge application for soil amendment (Hospido et al. 2012). Kalbar et al. (2012) conducted a multi-criteria decision analysis (MCDA) for wastewater treatment technologies for high-rise buildings in urban India. Two LCA criteria, eutrophication and global warming potential (GWP), and one Life Cycle Costing (LCC) criteria, Net Present Worth, were included along with other five. Compared to SBR and CAS, MBR was the preferred alternative. Its lower land and workforce requirements, higher reliability, and effluent quality offset the higher capital, operation and maintenance costs. Despite the common trends, the absolute impacts were different according to the selected configuration and operating conditions. Pretel et al. (2013) analyzed the environmental impacts of submerged anaerobic MBR systems operating at different temperatures. It was noted that in these systems it was important to maximize nutrient recovery as eutrophication could be reduced by up to 50% beyond the improvement in the categories of ecotoxicity and aquatic toxicity. More recently, Ioannou-Ttofa et al. (2016) focused its environmental impact analysis on an MBR wastewater facility (including construction and operation) operated with the target of antibiotic removal. The study concluded that the construction phase had minimal environmental impact compared to the operating phase where energy consumption was the main impact. Holloway et al. (2016) evaluated two technologies for water reuse i) an advanced bioreactor approach (FAT) ii) a hybrid ultrafiltration osmotic membrane bioreactor (UFO+MBR). It was concluded that the environmental impacts associated with FAT technology were lower than UFO+MBR technology due to the energy consumption of this technology.

Moreover, MBRs have also been operated under anaerobic conditions (AnMBR). Smith et al. (2014) compared the environmental and economic profile with another promising alternative: high rate activated sludge with anaerobic digestion (HRAS+AD) with more established counterparts: AeMBR+AD and CAS+AD. For medium strength waters, HRAS+AD was the only technology with a positive net energy balance. Despite the fact that the AnMBR recovers energy, it also uses 4 times more energy. In high strength water treatment, AnMBR achieved a more positive net energy balance than HRAS+AD. Although both technologies recovered more energy under these conditions, HRAS also required more energy for aeration, while the demand for AnMBR remained almost identical. As a result, AnMBR had higher environmental impacts than HRAS in all categories evaluated. In the case of GWP, it was due to the presence of dissolved methane in the effluent, while in the other categories it was due to its higher energy demand.

Pretel et al. (2015) evaluated the AnMBR from an operational, environmental and economic perspective. Different scenarios were studied for operating parameters such as sludge concentration, SRT, flow, mixed liquor suspended solid (MLSS) and specific gas demand (SGD) per area. Flux and MLSS were the two parameters that most affected LCA outcomes. Moreover, they concluded that submerged AnMBR could be a net positive energy and contribute to the development of other wastewater treatment technologies. In another study, Pretel et al. (2016) compared AnMBR with other aerobic wastewater treatment technologies. AnMBR was combined with three different post-treatments: (i) a conventional aerobic (ii) CAS and (iii) MBR. They concluded that AnMBR+CAS was the most sustainable option. In addition, this treatment could achieve a reduction in all impact categories.

Shoener et al. (2016) analyzed 150 AnMBR configurations. These differed in the type of reactor; the configuration, type and material of membranes; and the recovery and use of methane. The suitability of the membrane as well as the use of granular activated carbon (GAC) mostly affected economic and environmental sustainability. A sensitive analysis conducted among the most affordable alternatives suggested that in order to improve the environmental performance of the AnMBR, research should prioritize the reduction of cross-flow rate, removal of gas sparging, decrease of upward flow rate for bed expansion, and development of low-cost media for fouling mitigation. No design had both the lowest costs and the lowest environmental impacts, although those whose costs were below the 10th percentile also had comparatively low environmental impacts. That is to say, two configurations: a) cross flow with multi-tube, and b) hollow submerged fiber with GAC were less expensive than the current alternatives. Finally, Becker et al. (2017) evaluated different wastewater treatment technologies applied to food waste. These technologies were: (i) HRAS (ii) CAS and (iii) AnMBR. In addition, HRAS was studied with three disposal options: (i) AD (ii) landfill (LF) and (iii) composting (CP). They concluded that the most interesting options were HRAS + AD and AnMBR due to the biogas production. However, the emissions associated with biogas losses increased and impacts in the climate change category were worse in these scenarios.

Beyond these impacts, the LCA methodology needs to address all potential impacts associated with water use (Kounina et al. 2013). The life cycle inventory only provides the volume of freshwater used with limited information on its origin (resource type) and none on its destination (volume, quality and point of discharge) (Bayart et al. 2010). The ReCiPe methodology is currently the only one that reports the "freshwater depletion" indicator, so that the total amount of water used is specified. In a previous paper, the ReCiPe methodology was applied to estimate the water depletion indicator in two different scenarios (Pintilie et al. 2016). The first configuration consisted of a conventional wastewater treatment plant that discharged the treated effluent into natural watercourses, while the second scenario included tertiary treatment, which resulted in a treated effluent of sufficient quality to be used as process water in industrial operations. One of the advantages observed for the second scenario was the environmental credits for the use of reclaimed water. When analyzing this category, it was observed that in the scenario with water reuse, a negative value was obtained in the water depletion category ($-0.44\ m^3$), which results in an environmental benefit compared to the scenario that does not implement water reuse ($5.74 \times 10^{-4}\ m^3$). Consequently, the pressure on freshwater consumption is considerably reduced.

As an alternative to the indicator of water depletion in the ReCiPe methodology, several alternatives were evaluated and developed to quantify water use on a regional or global scale. The World Business Council for Sustainability Development initiated the international working group called WULCA. They applied the life-cycle approach to estimate water use and the water scarcity footprint indicator (Kounina et al. 2013). WULCA addresses specific objectives from the development of a general framework for assessing water use, as well as methods to characterize water use and related impacts (Kounina et al. 2013). For the first time the potential impacts were integrated into LCA and the WTA (withdrawal-to-availability ratio) was used as a characterization factor (Frischknecht et al. 2016). Subsequently, the methods focused on the water consumption-to-availability (CTA) ratio were developed. These methods are based on the rationale that the amount of water withdrawn from the environment and released to the same watershed do not contribute to local water scarcity (Boulay et al. 2015). The WULCA group identified the need for a transition factor between the WTA and CTA. This factor was defined as DTA (demand-to-availability ratio). These concepts were grouped into a method called AWARE (Availability Water Remaining).

The aim of this research study is the environmental evaluation of a decentralized scheme that evaluates a Membrane Bioreactor (MBR) for the treatment of urban wastewater generated at the METU University Campus in Ankara (Turkey), in order to reuse reclaimed water for irrigation of green areas. Within this framework, special focus was paid to the category of water use according to the AWARE method.

The objective of this chapter is to establish a basic roadmap for decentralized wastewater treatment technology schemes specifically for membrane bioreactors under the perspective of water scarcity footprint. First, the wastewater treatment plant (WWTP) was described focusing on the MBR. Secondly, the WWTP was evaluated according to the life cycle assessment approach to identify the critical stages with high environmental impact. Finally, a comparison was made between centralized and decentralized systems, with the aim of presenting a global vision of these treatment systems.

2 Materials and Methods

2.1 *Case of Study Description and Operation of MBR Facility*

The system includes a submerged rotary membrane vacuum bioreactor (VRM) designed for a capacity of 2000 hab-eq (Komesli and Gökçay 2014). The plant consists of two tanks and peripheral equipment (Fig. 2). The two tanks are separated by a wall, which in turn connects the two tanks through five holes at the bottom of the wall that separates them. The volume of the first aerobic tank is 85 m^3. The second anoxic tank (23 m^3) is used to house the membrane unit. The wastewater from the residence halls and the university is collected in a 10 m^3 storage tank and pumped to the treatment plant. At the inlet of the aeration tank, a screw-type sieve separates wastewater from materials larger than 3 mm (Komesli et al. 2007, 2015).

In the membrane unit, its support operates at a rotation speed of 2.5 rpm. The treated wastewater is pumped to the membrane modules by means of six radial hoses (Komesli et al. 2015). The membrane unit operates in intermittent vacuum cycles. The aeration tank is equipped with several membrane-type diffusers located at the bottom of the tank. The sludge is partially recirculated from the membrane tank at a variable frequency to control the concentration of MLSS (Komesli et al. 2007). Polyethersulfone membrane (PES) modules are flat type with a pore size of 0.038 μm and a total surface area of 540 m^2. There is a membrane support driven by an electric motor that provides the rotation speed to the filter support unit, creating a cross-flow

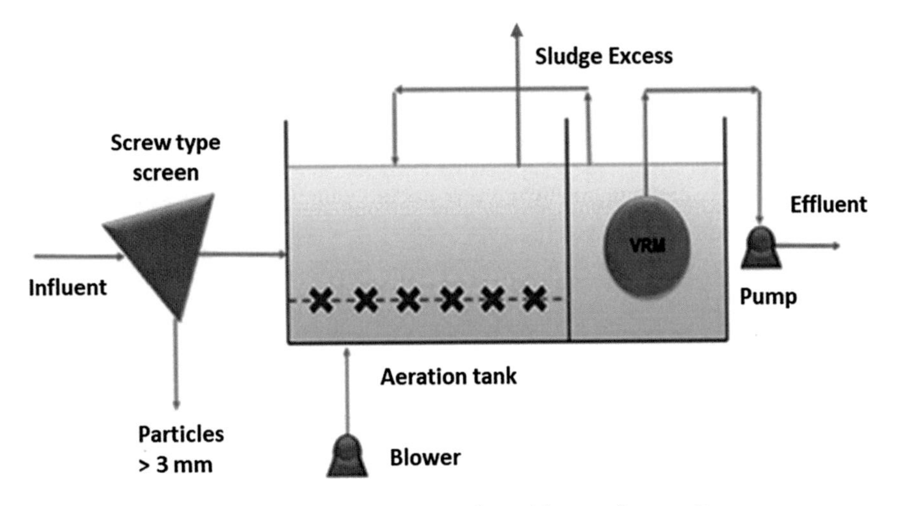

Fig. 2 Flow diagram of the wastewater treatment plant with a membrane unit

Table 3 Operational parameters of the MBR

Operating time (years)	10
Membrane	HUBER, A.G-Flat type
Flow (m^3/d)	108–172.8
Permeate flow (L/h·m^2)	8.3–13
HRT (h)	15–24
SRT (d)	10–150
TMP (mbar)	−80–(−300)
Organic load (kg/m^3·d)	0.31–1.53
Nitrogen charge (kg/m^3·d)	35–70

on the surface of the membrane modules along with a coarse aeration from the bottom of the modules (Komesli et al. 2007).

The pumping system operated in periods of 10 min (8 min vacuum and 2 min relaxation). The operational regime of the pump was changed because the MBR worked equally well at lower and more frequent vacuum periods; therefore, the cycle was changed to 5 min cycles (4 min vacuum and 1 min relaxation). During the relaxation time, the vacuum pump stopped but aeration and rotation continued (Komesli et al. 2015). Table 3 shows the operating conditions of the reactor during 10 years of operation.

The maintenance of the membrane unit is carried out with chemical cleaning when the intermembrane spaces are clogged, resulting in decreased permeate flow and increased transmembrane pressure (TMP), especially when the TMP remains constant and below −300 mbar. The membrane modules are chemically cleaned with 0.5% NaClO for 5–10 h, two or three times a year. During the 10 years of operation, the membrane was not changed or replaced. However, in the 5th year of operation, the damaged membrane surface area was about 50 m^2, but the flow rate increased to reach the same flow rate as in the first years of operation (Komesli et al. 2015). Moreover, this study includes a sensitivity analysis with different solids retention times (SRTs), which was a parameter that was modified during the operation phase. It is therefore necessary to know whether or not this parameter affects environmental impacts.

2.2 Functional Unit

The functional unit (FU) is defined as the evaluation of any product or process and its objective is to provide a reference of the related inputs and outputs of the system (ISO 14040 2006). Based on studies that consider the reuse of treated wastewater (Hospido et al. 2012; Lorenzo-Toja et al. 2016a; Pasqualino et al. 2011), the functional unit is defined as 1 m^3 of treated wastewater.

2.3 System Boundaries and Description

The selection of system boundaries is a key step in determining the unit processes included in the assessment (Lundin et al. 2000). Previous studies of conventional treatment plants have shown that the environmental impacts of the construction and decommissioning phases are lower than those of operation and maintenance, about 25–35% (Lorenzo-Toja et al. 2016b) or even negligible (Foley et al. 2010).

The current scheme includes the construction phase of the plant, since it can be an important factor in decentralized systems. The system depicted in Fig. 2 includes the construction, operation and maintenance phases of the MBR, as well as the treatment of sludge and the reuse of treated water for irrigation of the green areas of the university campus. Figure 3 also shows that during the construction phase, the production of blowers, pumps and pipes is outside the system boundaries due to the lack of data. If considered, this would cause a great deal of uncertainty for the inventory at this stage. Also, within the operational phase of the plant, wastewater collection and screening systems for solids removal are excluded.

2.4 Life Cycle Inventory, Data Quality and Simplifications

Inventory analysis involves the collection of data and calculation procedure to quantify relevant inputs and outputs of the system. The quality of the data used determines the interpretation of the environmental results and, consequently, the improvement actions to be taken into account. Therefore, the optimal solution is to have a wide

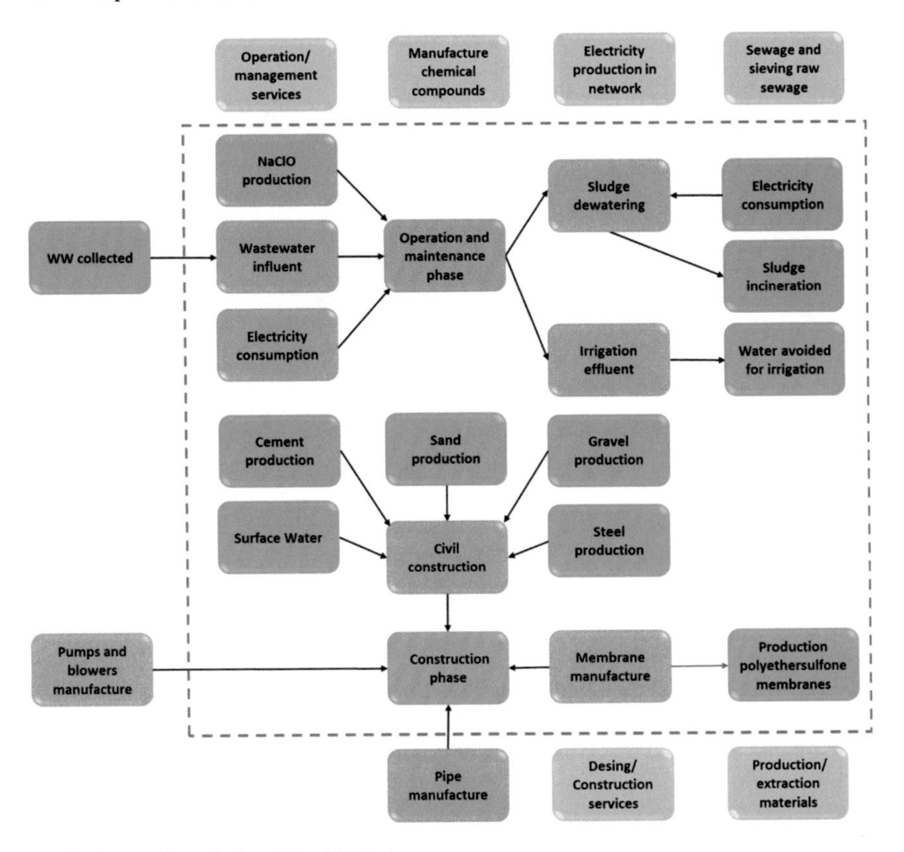

Fig. 3 System boundaries of the MBR plant

knowledge of the system, as well as real data with the minimum assumptions to obtain the most correct evaluation (ISO 14040 2006).

The primary and secondary data are presented in Table 4. Primary data correspond to the operation of the facility for 10 years. Secondary data were obtained from the Ecoinvent v3.3 database (Weidema et al. 2013). The simplifications considered in this work are presented below. Euro 4 Trucks (3.5–7.5 t) were selected as transport vehicles for chemicals and sludge, due to the smaller amount generated in the decentralized systems, in contrast to those of conventional wastewater treatment plants, between 16 to 32 t (Lorenzo-Toja et al. 2016a). As for chemicals, NaClO was obtained from the Ecoinvent v3.3 database. The polyelectrolyte doses required for sludge dewatering were those of typical ranges: 5–8 kg polymer/1000 kg dry matter (Tchobanoglous et al. 1998). The amount of sodium hypochlorite was calculated from Komesli et al. (2015). There were two membrane cleaning per year. The electricity production in Turkey was updated for 2016. High voltage electricity was converted to medium voltage considering atmospheric emissions and losses in transport (Dones et al. 2007).

Moreover, reports and construction projects were retrieved to complete missing data. Data on the construction of the facility and the membrane are presented in Tables 5 and 6, respectively. The data obtained for the construction of the membrane are all primary data except the thickness (Judd 2011) that was estimated according to the information provided by the manufacturer (HUBBER). The density of the polymer is taken from the wolfram-alpha database.

2.5 *Assessment Methodology and Impact Categories*

The methodologies used for the assessment of life-cycle impacts were AWARE (Availability Water Remaining), to determine the impact of water reuse for irrigation and water scarcity footprint, and CML (Guinée 2002) and IPCC methodologies (IPCC 2013) for the eutrophication and climate change categories, respectively. SimaPro v.8.2 software was used for the impact assessment.

The AWARE method calculates a mid-point indicator of water use, which indicates the available remaining water in a watershed relative to the world average, after meeting the demands of humans and aquatic ecosystems (Frischknecht et al. 2016). To assess potential water deprivation, it is assumed that the lower amount of available water per area, the greater the probability of depriving another user of this resource (Kounina et al. 2013). The method is based on the difference between availability and demand (1/AMD). When the demand is equal to or greater than the availability (negative value of AMD), CF_{AWARE} is set as a maximum (Puerto 2013). The AWARE category is represented by Eqs. 1 and 2.

$$AMD_i = \frac{(\text{Availability} - \text{HWC} - \text{EWR})}{\text{Area}} \quad (1)$$

$$CF_{AWARE} = \frac{AMD_{global\_average}}{AMD}, \text{ for Demand} < Availability \quad (2)$$

The HWC refers to the sum of human water consumption and the EWR refers to environmental water requirements. This methodology is used to determine the water scarcity footprint and to evaluate the water deprivation potential of other users when they consume water in a given geographical area. The first step is to calculate the Availability Minus Demand (AMD) of humans and aquatic ecosystems related to a given area, expressed in $m^3/(m^2 \cdot \text{month})$. Secondly, the value is normalized with the world average result ($AMD = 0.0136\ m^3/m^2 \cdot \text{month}$) and inverted; this value represents the relative value compared to the annual average of water consumption (in m^3). The world average is calculated as a consumption-weighted average. This indicator ranges from 0.1 to 100. Value 1 corresponds to an area with the same amount of available water. In contrast, the value of 10 represents an area where there is 10 times less water (Frischknecht et al. 2016). The inverse (1/AMD) represented the surface-time equivalent to generate one cubic of meter of unused water in the region under study (Frischknecht et al. 2016).

Table 4 Life cycle inventory (LCI) of MBR plant per 1 m^3 of treated wastewater

Inputs		Outputs	
From the technosphere		To the environment	
Materials and fuel		**Emissions to water**	
Influent		COD (g)	15.23
COD (g)	445.23	NH_4 (g)	18.47
NH_4 (g)	54.59	NO_3 (g)	3.04
NO_3 (g)	0.17	PO_4^{3-} (g)	4.68
PO_4^{3-} (g)	8.24		
Electricity consumption		**To the technosphere**	
Aerator (kWh)	1.40	**Products and co-products**	
MBR (kWh)	1.07	Sludge to incineration (g)	840.49
Centrifuge dehydrator[c] (kWh)	$2.08 \cdot 10^{-3}$		
Chemical consumption			
NaClO (15%)[a] (g)	3.69		
Polyelectrolyte[b] (g)	0.3		
Transport			
NaClO (15%) (kg·km)	$9.23 \cdot 10^{-2}$		
Sludge[b] (kg·km)	45.93		
Construction			
Infrastructure[d]	$7.96 \cdot 10^{-7}$		
PES membrane[a] (g)	0.32		

[a]Komesli et al. (2015)
[b]Tchobanoglous et al. (1998)
[c] Elias (2012)
[d]Doka (2003)

Table 5 Characteristics of membrane unit

Material	Polyethersulfone (PES)
Surface area (m^2)	540
Pore size (μm)	0.038
Thickness (μm)	300
Material density (kg/m^3)	1470
Membrane weight (kg)	238.14

Table 6 Inventory data for wastewater treatment plant construction for 2,000 hab-eq

Material/Construction process	Units	Value
Excavation by hydraulic digger	m^3	5,512
Transport by lorry	t·km	78,070
Transport by train	t·km	92,341
Electricity consumption by construction	kW·h	63
Concrete	m^3	1,584
Reinforcing steel	kg	122,879
Tap consumption	kg	193,204
Aluminium	kg	1,378
Limestone	kg	33,975
Stainless steel	kg	9,868
Fiberglass	kg	3,104
Copper	kg	1,457
Synthetic rubber (EPDM)	kg	1,394
Rock wool (insulation material)	kg	1,378
Organic chemical compounds	kg	6,415
Bitumen	kg	792
Inorganic chemical compounds	kg	792
Low density polyethylene (LDPE)	kg	32
High density polyethylene (HDPE)	kg	3,865
Polyethylene terephthalate (PET)	kg	3,896

Water availability represents renewable water; the values are taken by the Water-Gap model that is an average model for a period of 50 years. This model includes human consumption estimated for different sectors such as domestic, agricultural, livestock, among others. Ecosystem demand is evaluated using the variable flow method (VFM). This method classifies the flow regime as high, medium or low. The annual variability is taken into account to preserve aquatic ecosystems (Schenker et al. 2015).

For the interpretation of the results with respect to the world average, it is important to understand that a characterization factor of 1 is not equivalent to the factor for the average water consumption in the world. That is, the factor we can use when the location is unknown. This value is calculated as the weighted average of the factor, based on 1/AMD and not AMD. This implies that the consumption has a value of 43 for unknown uses and 20 and 46 for non-agricultural and agricultural water consumption, respectively (Boulay et al. 2015). The water scarcity footprint can be calculated by means of the AWARE methodology as the product of water consumption and the characterization factor as shown in Eq. (3).

Table 7 LCA results for the impact categories under assessment. FU: 1 m^3 of treated wastewater

Process	IPCC	Eutrophication	AWARE
Sub-processes	kg CO_2 eq/m^3	g PO_4^{3-} eq/m^3	m^3 world eq
NaClO (15%)	$5.76 \cdot 10^{-4}$	$1.44 \cdot 10^{-3}$	$5.65 \cdot 10^{-4}$
Polyelectrolyte	$5.01 \cdot 10^{-4}$	$4.56 \cdot 10^{-4}$	$3.07 \cdot 10^{-4}$
NaClO Transport	$4.76 \cdot 10^{-5}$	$4.33 \cdot 10^{-5}$	$5.84 \cdot 10^{-6}$
Sludge Transport	$2.37 \cdot 10^{-2}$	$2.15 \cdot 10^{-2}$	$2.90 \cdot 10^{-3}$
Plant Infrastructure	0.18	0.33	1.56
Membrane construction	$3.01 \cdot 10^{-3}$	$4.67 \cdot 10^{-3}$	$4.09 \cdot 10^{-3}$
Aeration electricity	0.81	1.00	1.36
VRM electricity	0.62	0.76	1.04
Centrifuge electricity	$1.20 \cdot 10^{-3}$	$1.48 \cdot 10^{-3}$	$2.03 \cdot 10^{-3}$
Irrigation	−0.39	−0.69	−55.81
Incineration	$3.35 \cdot 10^{-2}$	0.21	$2.08 \cdot 10^{-2}$
Influent	–	11.41	–
Effluent	–	36.07	–
Reuse	1.28	13.77	−51.81
No reuse	2.08	14.46	7.51

$$\text{Water scarcity footprint} = \text{water consumption} \cdot \frac{1}{\text{AMD}} = m^3_{\text{global\_average}} \tag{3}$$

Eutrophication was calculated using the CML methodology (Guinée 2002). This method takes into account the impact attributed to COD. In WWTPs, COD is a limiting discharge parameter (91/271/CCE) and in recent methods, such as ReCiPe (Goedkoop et al. 2009), this parameter is not included. For this reason, the CML method was selected.

The most updated characterization factors for the climate change category are found in the IPCC method; therefore, this method is more appropriate for calculating the impacts associated with this category (IPCC 2013).

3 Results

The plant data were analyzed for the three selected environmental categories (climate change, eutrophication and AWARE). The results were estimated on the basis of the functional unit that is one cubic of treated wastewater (Table 7). The following sections will explain in more detail the impacts for the different categories.

3.1 Climate Change (CC)

The total impact obtained in this category is 1.28 kg CO_2 eq/m^3 of treated wastewater, considering that the treated effluent is reused to irrigate the green areas of the university campus. Water reuse results in 23.8% less emissions compared to irrigation with tap water, which involves higher emissions in the climate change category: 2.08 kg CO_2 eq/m^3.

Taking into account the sub-processes of the examined system, the electrical consumption of the plant has the greatest environmental impact, with a contribution of 85.6%. The electricity consumption of the wastewater treatment is mainly attributed to the aerators of the biological reactor (0.81 kg CO_2 eq/m^3) and the vacuum, aeration and rotation of the Membrane Bioreactor, which accounted for 0.62 kg CO_2 eq/m^3. These emissions exceed the respective ones associated with the construction phase, and operationally the reuse of water shows a value of −0.39 kg CO_2 eq/m^3 to offset the impact of electricity consumption. The high contribution of energy consumption is related to the indirect emissions of non-biogenic CO_2 from fossil resources, with a share of 66.2% in the profile of energy production in Turkey.

Impacts related to the consumption and transport of chemicals, as well as to the transport of sludge, can be considered negligible. Chemical consumption (NaClO and polyelectrolyte) represents 0.08% of the total impacts. The transport of sludge and chemicals is also minimal, with 2% of the total impacts.

3.2 Eutrophication Potential (EP)

MBR technology is capable of reducing the EP by 68% when the effluent is compared between the treated effluent (11.4 g PO_4^{3-}/m^3) or untreated effluent (36.1 g PO_4^{3-}/m^3). The operation (1.76 g PO_4^{3-}/m^3), the construction of the plant (0.33 g PO_4^{3-}/m^3) and the incineration of sludge (0.21 g PO_4^{3-}/m^3) are the main sub-processes that contribute most to energy consumption, followed by the discharge of the treated effluent into the environment. The impacts associated with electricity consumption are related to the indirect emissions from coal mining to produce energy. The same processes affect the construction phase, since it is an energy-intensive process.

Furthermore, the reuse of treated water represents an environmental benefit in this category with a contribution of 4.8%. The latter is compensated for the impacts generated during the production and transport of chemicals, the manufacture of the membrane and the processes associated with sludge management, which accounts for 3% of the impacts.

3.3 Availability Water Remaining (AWARE)

In this category, as in the other impact categories, the construction of plant infrastructure (1.56 m^3 world eq.) and electrical consumption during plant operation (2.40 m^3 world eq.) are the stages that imply higher water consumption and, therefore, the largest contributors to the water scarcity footprint (Table 6). The greatest potential for water deprivation for other users is derived from the construction process and energy used to operate the plant. On the other hand, the reuse of treated water for irrigation led to a negative water scarcity footprint. The latter indicates an avoided deprivation from other users of 51.81 m^3 of water per m^3 of treated wastewater. This result is due to the fact that, when water is reused, that amount of water is avoided to be drawn from the water network. Therefore, the capture, distribution and treatment stages are avoided. If water reuse would not be taken into account, the water scarcity footprint would increase up to 7.50 m^3 of water per m^3 of treated wastewater.

For the climate change category, the impacts associated with transport and consumption of chemicals can be considered negligible. In this case, NaClO and polyelectrolyte represent 0.001 m^3 world eq./m^3 of treated wastewater, while transport accounts for a higher value of 0.03 m^3 world eq./m^3 of treated wastewater.

4 Discussion

4.1 Trade-off Analysis of Climate Change and Eutrophication Impact Categories

Despite its major relevance in environmental awareness, climate change is not the most important category when a WWTP is evaluated from an environmental point of view (Larsen et al. 2007). The impacts caused by the direct or indirect emissions associated with greenhouse gases (GHG) or other activities have a crucial effect on air pollution (IPCC 2013).

In the MBR, the impacts are caused by the energy consumption associated with the operational phase, in agreement with other papers (Hospido et al. 2012; Ioannou-Ttofa et al. 2016), who reported that energy consumption for the plant is responsible for more than 95% of the impacts. The construction phase accounts for 28.7%, which is dependent on the use of concrete and steel for infrastructure, acquiring an importance that has been traditionally considered negligible.

When it comes to the identification of the most important category in the environmental analysis of a WWTP, eutrophication ranks first (Rodriguez-Garcia et al. 2011; Zang et al. 2015), due to the high nutrient content (nitrogen and phosphorus) present in wastewater. The implementation of a nitrification-denitrification process results in 54–58% reduction of the eutrophication potential (Larsen et al. 2007).

In this study, the net environmental benefit (NEB) was calculated. The NEB analyzed the difference between the potential environmental impacts (PEI) caused

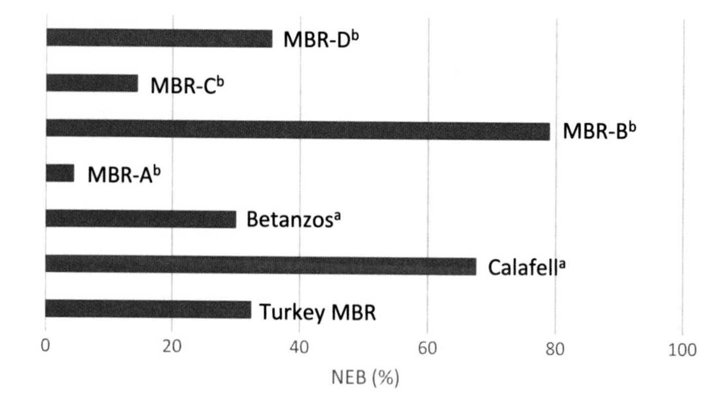

Fig. 4 Net environmental benefit comparison between different plants *Note* [a]Lorenzo-Toja et al. (2016a) [b] Hospido et al. (2012)

and avoided by the WWTP (Godin et al. 2012). PEI has been used in other studies to evaluate eutrophication in WWTPs (Lorenzo-Toja et al. 2016b; Rodriguez-Garcia et al. 2011). In this work, the NEB value was calculated and compared with other MBRs and conventional WWTPs (Fig. 4).

Figure 3 shows that the MBR facility has an average performance similar to that of a conventional WWTP (Lorenzo-Toja et al. 2016a). Compared to other studies on MBR (Hospido et al. 2012), the NEB value varies depending on the configuration. The NEB value of the MBR facility under study is similar to that of an UASB reactor followed by a hybrid reactor. The membrane is located in an independent chamber before the discharge of the effluent. The impacts associated with the treated effluent of an MBR vary from 13 to 20 g PO_4^{3-}/m^3. The results obtained in this study are within this range; thus, it can be concluded that MBR efficiency was maintained or even improved in the scale-up of the system.

4.2 *The Importance of Water Reuse: Giving the Floor to the AWARE Category*

This section will assess the water scarcity footprint, i.e. the water deprivation potential for other users, regardless of the type of user, in a given geographical area: Turkey. To date, no work has been published applying the AWARE methodology to assess the water scarcity footprint of wastewater treatment systems, only in the field of food production (Schenker et al. 2015; Bayart and Ekambi 2016). Several methodologies have been applied to determine the water footprint, such as ReCiPe, however, they have limitations. Opher and Friedler (2016) analyzed the impact of the water depletion in the centralized and decentralized systems. The results did not follow a general trend. This is due to the fact that the systems that convert seawater to drinking water have an environmental benefit in this category. The seawater is considered an infinite water source and therefore, its consumption has no impact. The centralized systems with no water reuse have less impact in this category than decentralized systems with water reuse, since they consider the consumption of drinking water which has impact in the water depletion category.

Morera et al. (2016) calculated the water footprint of the Garriga WWTP. The methodology used was the Water Footprint Network (WFN) methodology, which consists of classifying water into three types: Blue Water Footprint related to the water that evaporates during the operational phase of the WWTP, Grey Water Footprint associated with the concentration of the effluent and finally, Green Water Footprint related to the water evaporated by vegetation. In the case of WWTPs, green water is not considered. In this study, it is concluded that the water footprint is reduced by the secondary treatment, so there is a decrease in the grey water footprint when the wastewater is treated. In this study, this methodology is as inappropriate the water scarcity is not considered. This methodology was therefore not taken into account.

4.3 *Studying the Influence of SRT and the Construction Phase on the Environmental Outcomes*

A sensitivity analysis was conducted to identify the effect of solids retention times (SRTs) (10 to 140 days) on the climate change, eutrophication and AWARE indicators. Moreover, a comparative analysis was performed including and excluding the construction phase.

Influence of SRT on climate change category

As mentioned in Sect. 3.1, the main contributors to the climate change category in a decentralized MBR system are the construction phase and the operational energy consumption. Figure 5 shows that while SRT increases, energy consumption decreases, resulting in a lower impact in this category. The energy consumption has a deviation of ± 0.1 kWh/m^3, so the observed decrease is not relevant, which reflects that this variable does not depend on the SRT provided that the reactor operation is viable and efficient. On the other hand, the higher the SRT, the greater the amount of excess sludge and, as a consequence, the impacts due to the transport of sludge and the incineration process increase.

Figure 6 compares the overall impacts of all scenarios examined (with and without reuse of treated effluent) for the different sludge retention times. The impact of not reusing the water is lower at higher SRTs, (2.13 kg CO_2 eq/m^3 to 1.87 kg de CO_2 eq/m^3). This decrease is due to lower energy consumption during the operation. However, the decrease is not relevant due to the standard deviation of energy consumption, indicating the independence from the SRT.

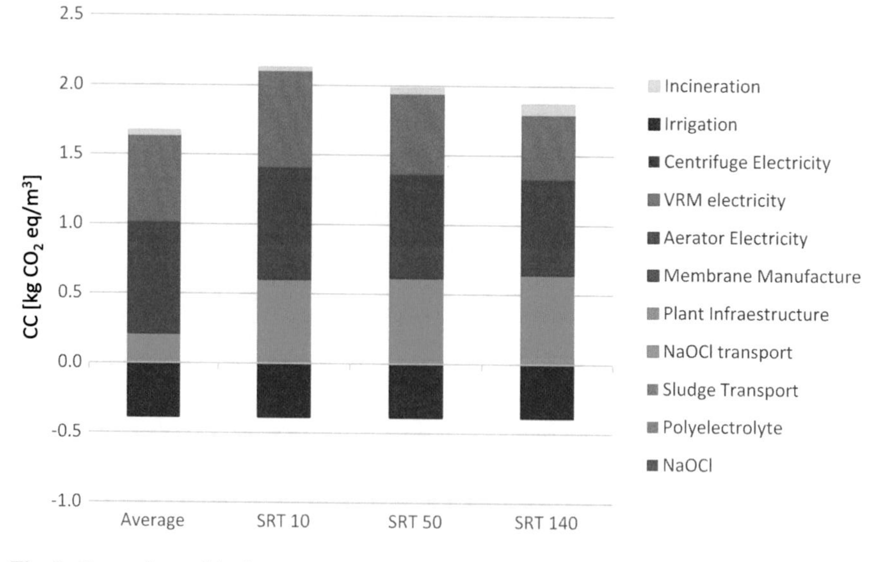

Fig. 5 Comparison of the impacts associated with the climate change category for different operating SRTs

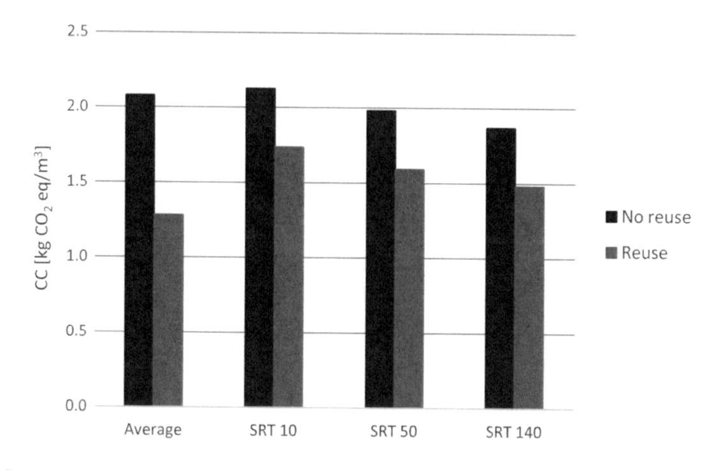

Fig. 6 Comparison of the climate change impacts according to different scenarios of water reuse

Influence of SRT on eutrophication category

The treated effluent is the main contributor to impacts in the eutrophication category. This is due to the content of organic matter, nitrogen and phosphorus. Figure 7 shows the variation in the eutrophication impact as a function of SRT. The MBR is able to reduce the impact of eutrophication by approximately 52–81%. The change at different SRT is that at low SRTs (i.e. 10 days), nitrification is not effective. The optimum reactor performance was achieved at an SRT of 50 days. However, when the

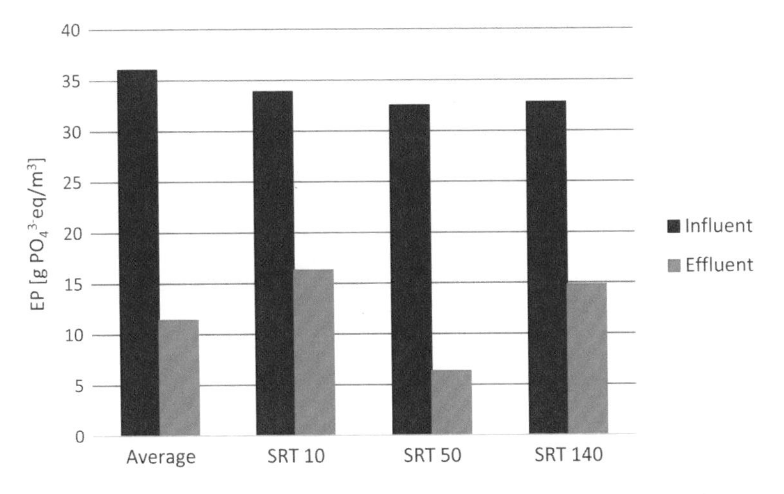

Fig. 7 Comparison of the eutrophication impacts according to different scenarios of water reuse

SRT is high (140 days), excessive biomass accumulation arises as a major operational problem.

The nutrient content of the effluent is the main contributor to eutrophication. The impact of the water effluent is similar to those of STR of 40 day and 140 day: 33% (Fig. 6). The minimum impacts of eutrophication are when the STR is 50 days, the impacts represent about 12%. Therefore, the optimal selection of SRT is based on the pollutant load of the effluent, but not exclusively on it. There are other important factors such as dissolved oxygen due to an inefficient operation of the aerators or an inadequate equipment sizing.

If eutrophication is analyzed by sub-processes (Fig. 8), the electricity consumption decreases by applying an SRT of 140 days; however, this reduction is negligible. Therefore, the sub-processes are independent of the SRT.

Influence of SRT on the AWARE category

The AWARE category allows the assessment of the water scarcity footprint. The reuse of irrigation water remains constant and independent of the solids retention time. Therefore, the reduction in energy consumption affects the category of water use for all SRTs examined (Fig. 9). In this case, the category is independent of the different SRTs, as was observed for the climate change category.

Influence of the construction phase on the impact categories

The construction phase has an impact on all categories, thus, an environmental analysis has been carried out without the construction phase. The results of the different categories are shown in Table 8. The impact of climate change decreases by 14% when the construction phase is not taken into account. Moreover, the water reuse to irrigate green areas offsets the emissions caused by electricity consumption by approximately 28%. The transport of chemicals and sludge has an impact of 4%.

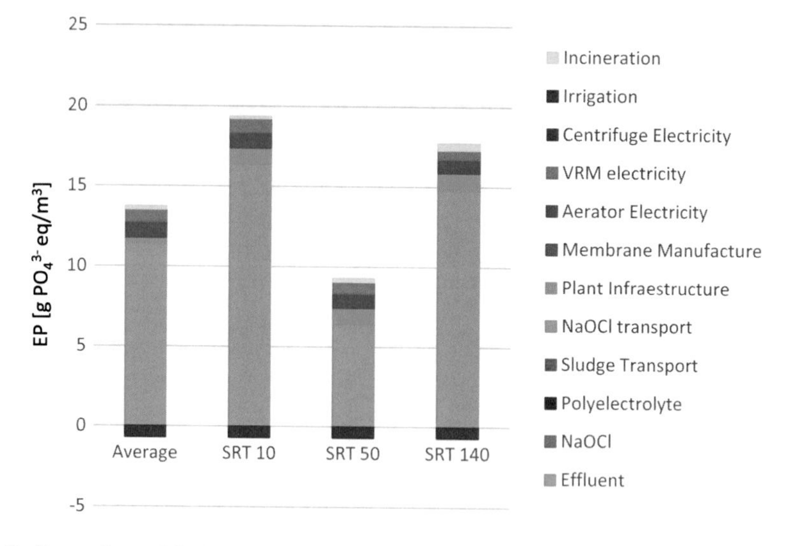

Fig. 8 Comparison of the impacts associated with the eutrophication potential for different operating SRTs

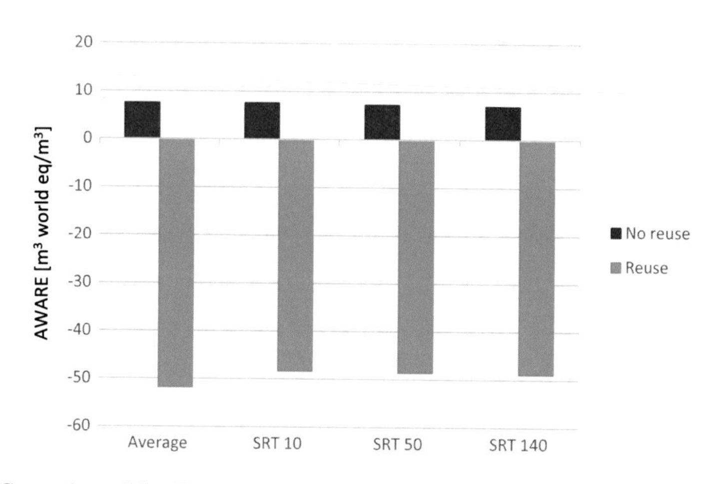

Fig. 9 Comparison of the AWARE impacts according to different scenarios of water reuse

The overall reduction in impacts is attributed to the contribution of emissions associated with the construction of tanks for the storage of the treated effluent. In the eutrophication category, the effluent remains the main source of impact (85%). The electricity consumption in the operation phase and sludge incineration are responsible for 15% of the impact. Water reuse offsets impacts in this category by 5%.

The evaluation of the impacts associated with water consumption, neglecting the construction phase, led to a considerable increase in the benefits of reuse. Conversely,

Table 8 LCA results without construction phase for the impact categories under assessment

Process	IPCC	EP	AWARE
Sub-processes	kg CO_2 eq/m^3	g PO_4^{3-} eq/m^3	m^3 world eq
NaClO (15%)	$5.76 \cdot 10^{-4}$	$1.44 \cdot 10^{-3}$	$5.65 \cdot 10^{-4}$
Polyelectrolyte	$5.01 \cdot 10^{-4}$	$4.56 \cdot 10^{-4}$	$3.07 \cdot 10^{-4}$
NaClO Transport	$4.76 \cdot 10^{-5}$	$4.33 \cdot 10^{-5}$	$5.84 \cdot 10^{-6}$
Sludge Transport	$2.37 \cdot 10^{-2}$	$2.15 \cdot 10^{-2}$	$2.90 \cdot 10^{-3}$
Membrane Construction	$3.01 \cdot 10^{-3}$	$4.67 \cdot 10^{-3}$	$4.09 \cdot 10^{-3}$
Aeration electricity	0.81	1.00	1.36
VRM electricity	0.62	0.76	1.04
Centrifuge electricity	$1.20 \cdot 10^{-3}$	$1.48 \cdot 10^{-3}$	$2.03 \cdot 10^{-3}$
Irrigation	−0.39	−0.69	−55.81
Incineration	$3.35 \cdot 10^{-2}$	0.21	$2.08 \cdot 10^{-2}$
Influent	–	11.41	–
Effluent	–	36.07	–
Reuse	1.10	12.77	−53.38
No reuse	1.40	13.46	2.43

if water is not reused for irrigation of green areas, the potential for water deprivation of other global users will increase.

4.4 *Evaluation of Sludge Management Strategy*

In this study, the final disposal of the sludge considers incineration as the existing alternative. This option is often not considered the best because of the high energy consumption and atmospheric emissions released (Suh and Rousseaux 2001). Accordingly, the alternative of its valorization in anaerobic digestion was considered to produce bioenergy and biofertilizers from the solid and liquid fractions of the digestate. The results for the three impact categories considered in this study are shown in Fig. 10.

The impacts on the CC category are high in the incineration process, approximately 23%. This is due to the energy consumption during this process. In addition, in the AD, biogas is reused in the plant, thus reducing the energy consumption from the grid. In the other impact categories, the trend is opposite to that of CC category. On the one hand, in the AWARE category, incineration has a better impact than AD because water is consumed in the composting process, therefore, water scarcity will increase when this process is used as a sludge management alternative. However, the difference between the two processes is not very high (10%). On the other hand, in

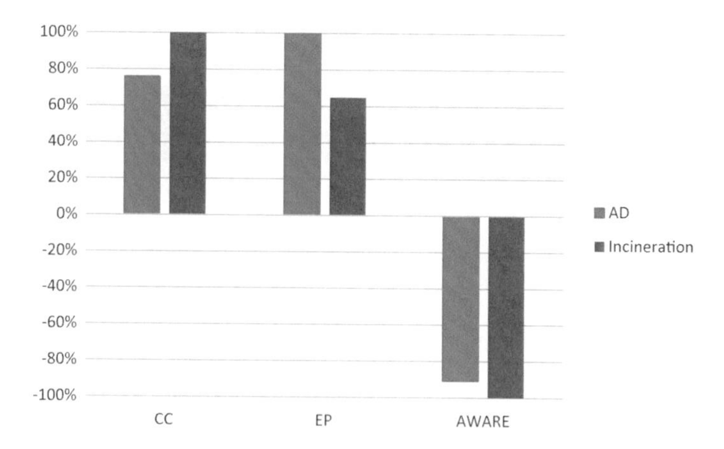

Fig. 10 Comparison between the AD sludge treatment and incineration sludge treatment

the EP category, the impacts are greater in the AD scenario. While the application of sludge to the soil as biofertilizer may be positive, the amount of nitrogen and phosphorus can also have negative effects on water resources. In the incineration process, the ashes are applied in the production of cement. In this case, the ashes were not transported to landfill, which would mean a greater environmental impact; on the contrary, the ashes that are considered a residual fraction can be exploited to obtain a valuable product.

There is no universal solution for sludge management, but it must be in line with the local application. The most effective methods involve energy consumption and reuse of materials, but these factors are not always possible. In the incineration process, valuable products such as nutrients are lost during the process. In addition, these methods are more expensive than others such as agricultural application. However, during this treatment it is necessary to control the leaching of heavy metals from the sludge.

4.5 *Broadening the Scope: Benchmarking Between Centralized and Decentralized Treatment*

Today, the boundaries of the WWTPs have been extended to include the entire treatment system. Traditionally, centralized systems have been used as the only alternative to guarantee urban hygiene (Opher and Friedler 2016), but the concept of wastewater treatment related to water reuse and nutrient recovery has gained importance in recent years (Lundie et al. 2004).

In this study, a modelled centralized system was compared with the decentralized system. The operation and construction phase, as well as the sewerage system were considered. The water reuse was only considered in the decentralized system. The

Table 9 Comparison between centralized and decentralized systems

Configuration	IPCC	EP	AWARE
	kg CO_2 eq/m^3	g PO_4^{3-} eq/m^3	m^3 world eq
Centralized	2.15	14.35	3.40
Decentralized	1.52	13.44	−51.70

impacts obtain for each impact category are shown in Table 9. Impact increases in centralized system for the studied categories. In the CC category, the impacts increase about a 17% for the centralized system. These impacts are associated with the electricity consumption in the construction of the sewer system that needs to collect wastewater form distant locations and the indirect emissions related to the diesel consumption by the machinery.

The EP category is not dependent on electricity consumption. The main impacts in this category are attributed to phosphorus, nitrogen and chemical oxygen demand. The main compound that generates impact is ammonia in water. This is because in the centralized system the effluent is discharged into the river, which is a sensitive area. In the decentralized system scenario, water is reused and not discharged into a sensitive area. Therefore, the impacts are lower than a centralized system. Finally, in the AWARE category, irrigation of green areas was not considered, so impacts are greater due to the deprivation of other water users and ecosystems. The main impacts are associated to indirect emission provoked in the water consumption during the construction phase.

5 Conclusions

The construction phase of a decentralized plant that applies the Membrane Bioreactor technology plays an important role in terms of the impacts associated with the target categories of climate change, eutrophication potential and water scarcity footprint. The latter were generally considered negligible in accounting for the impacts associated with energy consumption during the operation of an MBR.

The reuse of treated water significantly improves the environmental profile of the system in terms of the impact categories examined: 23.8% for climate change, 4.8% for eutrophication potential and 133.8% for the AWARE indicator. The construction phase has a negative impact on the decentralized system. However, the use of the MBR at decentralized scale allows for the reuse of treated water and improves the water potential of a country at risk of water scarcity.

Finally, the sensitivity analysis for the operation of the system at different SRTs showed that this parameter is independent in the different sub-processes. The eutrophication potential category is the only one that depends on the nutrient content of the effluent and is therefore affected by SRT. The optimum SRT of 50 days will imply a reduction of eutrophication impacts.

The main weaknesses of LCA studies are attributed to the collection and validity of the data required for LCI. This stage requires data on the consumption of chemicals, electricity, raw materials and water for each sub-process, as well as emissions to water, soil or air. In addition, the treatment selected for sludge is incineration, which can be very costly in terms of electricity consumption, and emissions to the atmosphere must be treated before their release. Incineration may not be the best option for treating sludge, but there is no universal solution for its management. In the future, it would be interesting to compare how the impacts of the plant change if it is elsewhere. Future research will focus on the advantages and disadvantages of the decentralized approach, as well as its possible application and integration into conventional treatment systems.

Acknowledgements This research was supported by the UE projects: Pioneer_STP (PCIN-2015-22 (MINECO)/ID199 (WaterJPI) and Run4Life (730285-1). The authors (A. Arias, G. Feijoo and M. T. Moreira) belong to the Galician Competitive Research Group (GRC ED431C 2017/29) and to the CRETUS Strategic Partnership (AGRUP2015/02).

References

Arslan-Alaton, I., Türkman, A., & Orhon, D. (2010). Future of water resources and wastewater reuse in Turkey. Chapter 31. In *Climate Change and Its Effects on Water Resources*. Springer.

Battistoni, P., Fatone, F., Bolzonella, D., & Pavan, P. (2006). Full scale application of the coupled alternate cycles-membrane bioreactor (AC-MBR) process for wastewater reclamation and reuse. IWA World Water Congr. Exhib.

Bayart, J.-B., & Ekambi, E. (2016). WULCA-AWARE case study: Application of AWARE to bottled water and beverages 28.

Bayart, J. B., Bulle, C., Deschênes, L., Margni, M., Pfister, S., Vince, F., et al. (2010). A framework for assessing off-stream freshwater use in LCA. *International Journal of Life Cycle Assessment, 15,* 439–453. https://doi.org/10.1007/s11367-010-0172-7.

Becker, A. M., Yu, K., Stadler, L. B., & Smith, A. L. (2017). Co-management of domestic wastewater and food waste: A life cycle comparison of alternative food waste diversion strategies. *Bioresource Technology, 223,* 131–140. https://doi.org/10.1016/j.biortech.2016.10.031.

Bixio, D., Thoeye, C., De Koning, J., Joksimovic, D., Savic, D., Wintgens, T., et al. (2006). Wastewater reuse in Europe. *Desalination, 187,* 89–101. https://doi.org/10.1016/j.desal.2005.04.070.

Boulay, A. M., Bare, J., De Camillis, C., Döll, P., Gassert, F., Gerten, D., Humbert, S., Inaba, A., Itsubo, N., Lemoine, Y., Margni, M., Mothoshita, M., Núñez, M., Pastor, A. V., Ridoutt, B., Schencker, U., Shirakawa, N., Vionnet, S., Worbe, S., Yoshikawa, S., & Pfister, S. (2015). Consensus building on the development of a stress-based indicator for LCA-based impact assessment of water consumption: Outcome of the expert workshops. *International Journal Life Cycle Assess.* 20.

Brepols, C., Dorgeloh, E., Frechen, F. B., Fuchs, W., Haider, S., Joss, A., et al. (2008). Upgrading and retrofitting of municipal wastewater treatment plants by means of membrane bioreactor (MBR) technology. *Desalination, 231,* 20–26. https://doi.org/10.1016/j.desal.2007.11.035.

Bui, X. T., Vo, T. P. T., Ngo, H. H., Guo, W. S., & Nguyen, T. T. (2016). Multicriteria assessment of advanced treatment technologies for micropollutants removal at large-scale applications. *Science of the Total Environment, 563–564,* 1050–1067. https://doi.org/10.1016/j.scitotenv.2016.04.191.

Cakmak, B., Uçar, Y., & Akuzum, T. (2007). Water resources management problems and solutions for Turkey. In *International Congress on River Basin Management* (pp. 867–880).

Chen, J. P., Mou, H., Wang, L. K., & Matsuura, T. (2006). Membrane filtration. In Wang, L. K., Hung, Y. -T., & Shammas, N. K. (Eds.), *Advanced Physicochemical Treatment Processes*. (pp. 203–259). Humana Press, Totowa, NJ. https://doi.org/10.1007/978-1-59745-029-4_7

Chen, R., & Wang, X. C. (2009). Cost-benefit evaluation of a decentralized water system for wastewater reuse and environmental protection. *Water Science and Technology, 59,* 1515–1522. https://doi.org/10.2166/wst.2009.156.

Coday, B. D., Xu, P., Beaudry, E. G., Herron, J., Lampi, K., Hancock, N. T., et al. (2014). The sweet spot of forward osmosis: Treatment of produced water, drilling wastewater, and other complex and difficult liquid streams. *Desalination, 333,* 23–35. https://doi.org/10.1016/j.desal.2013.11.014.

Comninellis, C., Kapalka, A., Malato, S., Parsons, S., Poulios, I., & Mantzavinos, D. (2008). Advanced oxidation process for water treatment: Advances and trends for R&D. *Journal of Chemical Technology and Biotechnology, 83,* 769–776.

Doka, G. (2003). *Life cycle inventories of waste treatment services*. Ecoinvent Report no. 13 Part IV: wastewater treatment. Swiss Centre for Life Cycle Inventories.

Dominguez, S., Laso, J., Margallo, M., Aldaco, R., Rivero, M. J., Irabien, Á., et al. (2018). LCA of greywater management within a water circular economy restorative thinking framework. *Science of the Total Environment, 621,* 1047–1056. https://doi.org/10.1016/j.scitotenv.2017.10.122.

Dones R., Bauer C., Bolliger R., Burger B., Faist Emmenegger M., Frischknecht R., Heck T., Jungbluth N., Röder A., & T. M. (2007). Life cycle inventories of energy systems: results of current systems in Switzerland and other UCTE countries. Ecoinvent Rep. 5. https://doi.org/10.1007/s11367-014-0838-7.

DSI, W. (2012). General directorate of state hydarulic works. *Ministry of Forestry adn Water Affairs*.

ECC. (1991). Directive 91/271/1991 ECC of 21 May 1991 concerning urban waste water treatment. *Official Journal of the European Communities*.

Elias, X. (2012). *Reciclaje de residuos industriales*. Díaz de Santos: Residuos sólidos urbanos y fangos de depuradora.

FAO. (2018). AQUASTAT [WWW Document]. Retrieved December 6, 2018, from, www.fao.org.

Fenu, A., Roels, J., Wambecq, T., de Gussem, K., Thoeye, C., de Gueldre, G., et al. (2010). Energy audit of a full scale MBR system. *Desalination, 262,* 121–128. https://doi.org/10.1016/j.desal.2010.05.057.

Foley, J., de Haas, D., Hartley, K., & Lant, P. (2010). Comprehensive life cycle inventories of alternative wastewater treatment systems. *Water Research, 44,* 1654–1666. https://doi.org/10.1016/j.watres.2009.11.031.

Frischknecht, R., Fantke, P., Tschümperlin, L., Niero, M., Antón, A., Bare, J., Boulay, A. M., Cherubini, F., Hauschild, M. Z., Henderson, A., Levasseur, A., McKone, T. E., Michelsen, O., i Canals, L. M., Pfister, S., Ridoutt, B., Rosenbaum, R. K., Verones, F., Vigon, B., & Jolliet, O. (2016). Global guidance on environmental life cycle impact assessment indicators: Progress and case study. *International Journal Life Cycle Assess. 21*, 429–442. https://doi.org/10.1007/s11367-015-1025-1.

Gil, M. J., Soto, A. M., Usma, J. I., & Gutiérrez, O. M. (2012).Contaminantes emergentes en aguas, efectos y posibles tratamientos. *Producción + limpia, 7*, 52–73.

Godin, D., Bouchard, C., & Vanrolleghem, P. A. (2012). Net environmental benefit: Introducing a new LCA approach on wastewater treatment systems. *Water Science and Technology, 65,* 1624–1631. https://doi.org/10.2166/wst.2012.056.

Goedkoop, M., Heijungs, R., Huijbregts, M., Schryver, A. De, Struijs, J., Zelm, R. Van (2009). ReCiPe 2008. Potentials 1–44. https://doi.org/10.029/2003JD004283

Guinée, J. (2002). Handbook on life cycle assessment operational guide to the iso standards. *Environmental Impact Assessment Review, 23,* 129–130. https://doi.org/10.1016/S0195-9255(02)00101-4.

Hochstrat, R., Joksimovic, D., Wintgens, T., Melin, T., & Savic, D. (2007). Economic considerations and decision support tool for wastewater reuse scheme planning. *Water Science and Technology, 56,* 175–182. https://doi.org/10.2166/wst.2007.570.

Høibe, L., Clauson-Kaas, J., Wenzel, H., Larsen, H. F., Jacobsen, B. N., & Dalgaard, O. (2008). Sustainability assessment of advanced wastewater treatment technologies. *Water Science and Technology, 58,* 963–968.

Holloway, R. W., Miller-Robbie, L., Patel, M., Stokes, J. R., Munakata-Marr, J., Dadakis, J., et al. (2016). Life-cycle assessment of two potable water reuse technologies: MF/RO/UV-AOP treatment and hybrid osmotic membrane bioreactors. *Journal of Membrane Science, 507,* 165–178. https://doi.org/10.1016/j.memsci.2016.01.045.

Hophmayer-Tokich, S. (2006). Wastewater management strategy: Centralized versus decentralized technologies for small communities. *CSTM-Reeks, 271,* 27.

Hospido, A., Sanchez, I., Rodriguez-Garcia, G., Iglesias, A., Buntner, D., Reif, R., et al. (2012). Are all membrane reactors equal from an environmental point of view? *Desalination, 285,* 263–270. https://doi.org/10.1016/j.desal.2011.10.011.

Ioannou-Ttofa, L., Foteinis, S., Chatzisymeon, E., & Fatta-Kassinos, D. (2016). The environmental footprint of a membrane bioreactor treatment process through Life Cycle Analysis. *Science of the Total Environment, 568,* 306–318. https://doi.org/10.1016/j.scitotenv.2016.06.032.

ISO 14040. (2006). International Organization for Standardization, ISO 14040. *Environmental management-life cycle assessment—principles and framework*, Geneve.

Jeong, H., Broesicke, O. A., Drew, B., & Crittenden, J. C. (2018). Life cycle assessment of small-scale greywater reclamation systems combined with conventional centralized water systems for the City of Atlanta. *Georgia Journal of Cleaner Production, 174,* 333–342. https://doi.org/10.1016/j.jclepro.2017.10.193.

Judd, S. (2011). Principles and applications of membrane bioreactors for water and wastewater treatment. The MBR Book.

Kalbar, P. P., Karmakar, S., & Asolekar, S. R. (2012). Technology assessment for wastewater treatment using multiple-attribute decision-making. *Technology in Society, 34,* 295–302. https://doi.org/10.1016/j.techsoc.2012.10.001.

Komesli, O. T., & Gökçay, C. F. (2014). Investigation of sludge viscosity and its effects on the performance of a vacuum rotation membrane bioreactor. *Environmental Technology (United Kingdom), 35,* 645–652. https://doi.org/10.1080/09593330.2013.840655.

Komesli, O. T., Muz, M., Ak, S., & Gökçay, C. F. (2015). Prolonged reuse of domestic wastewater after membrane bioreactor treatment. *Desalination and Water Treatment, 53,* 3295–3302. https://doi.org/10.1080/19443994.2014.934107.

Komesli, O. T., Teschner, K., Hegemann, W., & Gokcay, C. F. (2007). Vacuum membrane applications in domestic wastewater reuse. *Desalination, 215,* 22–28. https://doi.org/10.1016/j.desal.2006.10.025.

Kounina, A., Margni, M., Bayart, J.-B., Boulay, A.-M., Berger, M., Bulle, C., Frischknecht, R., Koehler, A., Milà i Canals, L., Motoshita, M., Núñez, M., Peters, G., Pfister, S., Ridoutt, B., van Zelm, R., Verones, F., & Humbert, S. (2013). Review of methods addressing freshwater use in life cycle inventory and impact assessment. *International Journal Life Cycle Assess. 18*, 707–721. https://doi.org/10.1007/s11367-012-0519-3.

Krzeminski, P., Langhorst, W., Schyns, P., de Vente, D., Van den Broeck, R., Smets, I. Y., et al. (2012). The optimal MBR configuration: Hybrid versus stand-alone—Comparison between three full-scale MBRs treating municipal wastewater. *Desalination, 284,* 341–348. https://doi.org/10.1016/j.desal.2011.10.038.

Lam, L., Kurisu, K., & Hanaki, K. (2015). Comparative environmental impacts of source-separation systems for domestic wastewater management in rural China. *Journal of Cleaner Production, 104,* 185–198. https://doi.org/10.1016/j.jclepro.2015.04.126.

Larsen, H. F., Hauschild, M., Wenzel, H., & Almemark, M. (2007). Homogeneous LCA methodology agreed by NEPTUNE and INNOWATECH. Denmark 1–34.

Lehtoranta, S., Vilpas, R., & Mattila, T. J. (2014). Comparison of carbon footprints and eutrophication impacts of rural on-site wastewater treatment plants in Finland. *Journal of Cleaner Production, 65,* 439–446. https://doi.org/10.1016/j.jclepro.2013.08.024.

Lorenzo-Toja, Y., Alfonsín, C., Amores, M. J., Aldea, X., Marin, D., Moreira, M. T., & Feijoo, G. (2016a). Beyond the conventional life cycle inventory in wastewater treatment plants. *Science of the Total Environment*, 553, 71–82. https://doi.org/10.1016/j.scitotenv.2016.02.073

Lorenzo-Toja, Y., Vázquez-Rowe, I., Amores, M. J., Termes-Rifé, M., Marín-Navarro, D., Moreira, M. T., & Feijoo, G. (2016b). Benchmarking wastewater treatment plants under an eco-efficiency perspective. *Science of the Total Environment*, 566–567, 468–479. https://doi.org/10.1016/j.scitotenv.2016.05.110

Lundie, S., Peters, G. M., & Beavis, P. C. (2004). Life cycle assessment for sustainable metropolitan water systems planning. *Environmental Science and Technology, 38,* 3465–3473. https://doi.org/10.1021/es034206m.

Lundin, M., Bengtsson, M., & Molander, S. (2000). Life cycle assessment of wastewater systems: Influence of system boundaries and scale on calculated environmental loads. *Environmental Science and Technology, 34,* 180–186. https://doi.org/10.1021/es990003f.

Luo, Y., Guo, W., Ngo, H. H., Nghiem, L. D., Hai, F. I., Zhang, J., et al. (2014). A review on the occurrence of micropollutants in the aquatic environment and their fate and removal during wastewater treatment. *Science of the Total Environment, 473–474,* 619–641. https://doi.org/10.1016/j.scitotenv.2013.12.065.

Massoud, M. A., Tarhini, A., & Nasr, J. A. (2009). Decentralized approaches to wastewater treatment and management: Applicability in developing countries. *Journal of Environmental Management, 90,* 652–659. https://doi.org/10.1016/j.jenvman.2008.07.001.

Matos, C., Pereira, S., Amorim, E. V., Bentes, I., & Briga-Sa´, A. (2014). Wastewater and greywater reuse on irrigation in centralized and decentralized systems—An integrated approach on water quality, energy consumption and CO2emissions. *Science of the Total Environment, 493,* 463–471. https://doi.org/10.1016/j.scitotenv.2014.05.129.

Memon, F. A., Zheng, Z., Butler, D., Shirley-Smith, C., Lui, S., Makropoulos, C., et al. (2007). Life cycle impact assessment of greywater recycling technologies for new developments. *Environmental Monitoring and Assessment, 129,* 27–35. https://doi.org/10.1007/s10661-006-9422-3.

Meuler, S., Paris, S., & Hackner, T. (2008). Membrane bio-reactors for decentralized wastewater treatment and reuse. *Water Science and Technology, 58,* 285–294. https://doi.org/10.2166/wst.2008.356.

Morera, S., Corominas, L., Poch, M., Aldaya, M. M., & Comas, J. (2016). Water footprint assessment in wastewater treatment plants. *Journal of Cleaner Production, 112,* 4741–4748. https://doi.org/10.1016/j.jclepro.2015.05.102.

Nogueira, R., Brito, A. G., Machado, A. P., Janknecht, P., Salas, J. J., Vera, L., et al. (2009). Economic and environmental assessment of small and decentralized wastewater treatment systems. *Desalination and Water Treatment, 4,* 16–21. https://doi.org/10.5004/dwt.2009.349.

Opher, T., & Friedler, E. (2016). Comparative LCA of decentralized wastewater treatment alternatives for non-potable urban reuse. *Journal of Environmental Management, 182,* 464–476. https://doi.org/10.1016/j.jenvman.2016.07.080.

Ortiz, M., Raluy, R. G., & Serra, L. (2007). Life cycle assessment of water treatment technologies: Wastewater and water-reuse in a small town. *Desalination, 204,* 121–131. https://doi.org/10.1016/j.desal.2006.04.026.

Pasqualino, J. C., Meneses, M., & Castells, F. (2011). Life cycle assessment of urban wastewater reclamation and reuse alternatives. *Journal of Industrial Ecology, 15,* 49–63. https://doi.org/10.1111/j.1530-9290.2010.00293.x.

Pintilie, L., Torres, C. M., Teodosiu, C., & Castells, F. (2016). Urban wastewater reclamation for industrial reuse: An LCA case study. *Journal of Cleaner Production, 139,* 1–14. https://doi.org/10.1016/j.jclepro.2016.07.209.

Pretel, R., Robles, A., Ruano, M. V., Seco, A., & Ferrer, J. (2016). Economic and environmental sustainability of submerged anaerobic MBR-based (AnMBR-based) technology as compared to aerobic-based technologies for moderate-/high-loaded urban wastewater treatment. *Journal of Environmental Management, 166,* 45–54. https://doi.org/10.1016/j.jenvman.2015.10.004.

Pretel, R., Robles, A., Ruano, M. V., Seco, A., & Ferrer, J. (2013). Environmental impact of submerged anaerobic MBR (SAnMBR) technology used to treat urban wastewater at different temperatures. *Bioresource Technology, 149,* 532–540. https://doi.org/10.1016/j.biortech.2013.09.060.

Pretel, R., Shoener, B. D., Ferrer, J., & Guest, J. S. (2015). Navigating environmental, economic, and technological trade-offs in the design and operation of submerged anaerobic membrane bioreactors (AnMBRs). *Water Research, 87,* 531–541. https://doi.org/10.1016/j.watres.2015.07.002.

Prieto, A. L., Vuono, D., Hollaway, R., Benecke, J., Henkel, J., Cath, T. Y., Reid, T., Johson, L., & Drewes, J. (2013). Decentralized wastewater treatment for distributed water reclamation and reuse: The good, the bad and the ugly. Experience from a case study, in: In *Novel Solutions to Water Pollution, American Chemical Society*. Chapter 15.

Puerto, M., 2013. Water scarcity footprint for cement production 1–4.

Remy, C., & Jekel, M. (2008). Sustainable wastewater management: Life Cycle Assessment of conventional and source-separating urban sanitation systems. *Water Science and Technology, 58,* 1555–1562. https://doi.org/10.2166/wst.2008.533.

Rodriguez-Garcia, G., Molinos-Senante, M., Hospido, A., Hernández-Sancho, F., Moreira, M. T., & Feijoo, G. (2011). Environmental and economic profile of six typologies of wastewater treatment plants. *Water Research, 45,* 5997–6010. https://doi.org/10.1016/j.watres.2011.08.053.

Schenker, U., Roy, P., Boulay, A., & Gaillard, B. (2015). Comparison of water scarcity methods : Meat & vegetarian burger case study 6.

Shehabi, A., Stokes, J. R., Horvath, A. (2012). Energy and air emission implications of a decentralized wastewater system. *Environmental Research Letters*, *7*. https://doi.org/10.1088/1748-9326/8/1/019001.

Shoener, B. D., Zhong, C., Greiner, A. D., O. Khunjar, W., Hong, P. -Y., & Guest, J. S. (2016). Design of anaerobic membrane bioreactors for the valorization of dilute organic carbon waste streams. *Energy and Environmental Science, 9,* 1102–1112. https://doi.org/10.1039/C5EE03715H

Smith, A. L., Stadler, L. B., Cao, L., Love, N. G., Raskin, L., & Skerlos, S. J. (2014). Navigating wastewater energy recovery strategies: A life cycle comparison of anaerobic membrane bioreactor and conventional treatment systems with anaerobic digestion. *Environmental Science and Technology, 48,* 5972–5981. https://doi.org/10.1021/es5006169.

Stocker, T. F., Qin, D., Plattner, G. K., Tignor, M., Allen, S. K., Boschung, J., Nauels, A., Xia, Y., Bex, B., & Midgley, B. M. (2013). IPCC, 2013: Summary for policymakers. In *Climate Change 2013: The Physical Science Basis. Contribution of Working Group I to the Fifth Assessment Report of the Intergovernmental Panel on Climate Change*.

Suh, Y., & Rousseaux, P. (2001). An LCA of alternative wastewter sludge treatment scenarios. *Resources, Conservation and Recycling, 35,* 191–200.

Tangsubkul, N., Beavis, P., Moore, S. J., Lundie, S., & Waite, T. D. (2005). Life cycle assessment of water recycling technology. *Water Resource Management, 19,* 521–537. https://doi.org/10.1007/s11269-005-5602-0.

Tangsubkul, N., Parameshwaran, K., Lundie, S., Fane, A. G., & Waite, T. D. (2006). Environmental life cycle assessment of the microfiltration process. *Journal of Membrane Science, 284,* 214–226. https://doi.org/10.1016/j.memsci.2006.07.047.

Tchobanoglous, G., Burton, F., & Stensel, H. D. (1998). Wastewater engineering: An overview. In *Wastewater Engineering Treatment and Reuse*, pp. 1–24. https://doi.org/10.1016/0309-1708(80)90067-6

Thibodeau, C., Monette, F., Bulle, C., & Glaus, M. (2014). Comparison of black water source-separation and conventional sanitation systems using life cycle assessment. *Journal of Cleaner Production, 67,* 45–57. https://doi.org/10.1016/j.jclepro.2013.12.012.

Wang, L. K., Chen, J. P., Hung, Y. T., & Shammas, N. (2011). Membrane and desalination technologies. In *Handbook of Environmental Engineering*. Humana Press.

Weidema, B. P., Bauer, C., Hischier, R., Nemecek, T., Reinhard, J., Vadenbo, C. O., & Wernet, G. (2013). *Overview and methodology. Data quality guideline for the ecoinvent database version 3*. Swiss Cent. Life Cycle Invent. St. Gall.

Wenzel, H., Larsen, H. F., Clauson-Kaas, J., Høibye, L., & Jacobsen, B. N. (2008). Weighing environmental advantages and disadvantages of advanced wastewater treatment of micro-pollutants using environmental life cycle assessment. *Water Science and Technology, 57,* 27–32. https://doi.org/10.2166/wst.2008.819.
Wiessmann, U., Choi, I. S., & Dombrowski, E. M. (2007). *Fundamental of biological wastewater treatment*. Wiley VCH.
Yuksel, I. (2015). Water management for sustainable and clean energy in Turkey. *Energy Reports, 1,* 129–133. https://doi.org/10.1016/j.egyr.2015.05.001.
Zang, Y., Li, Y., Wang, C., Zhang, W., & Xiong, W. (2015). Towards more accurate life cycle assessment of biological wastewater treatment plants: A review. *Journal of Cleaner Production, 107,* 676–692. https://doi.org/10.1016/j.jclepro.2015.05.060.